'Double Vision' artist: Tim Harrisson
Material: Carrara Marble
Date: 2004
Photographer: Jon Stone

Praise for *Jungian Analysts Working Across Cultures*

This fascinating compilation celebrates the expanding interest around the globe in the work of Jung and the clinical application of his ideas. Jan Wiener and Catherine Crowther are ideal candidates for the task of editing this excellent book. They are true pioneers and ambassadors of Jungian analysis. For over twenty-five years, they have shared their passion and enthusiasm for analytical psychology with countless trainee analysts in Russia and Eastern Europe but also further afield. Just as they do with their trainees, they manage to get the best out of their authors who give moving and stimulating accounts of their multi-cultural experiences. We see how orthodoxies are challenged and how Jungian analysis, like all things, is required to evolve to survive. I can't recommend it highly enough.

Martin Schmidt is a training analyst at the Society of Analytical Psychology, London and lecturer/supervisor in Jungian analysis in Russia, Ukraine, and Serbia.

This inspiring book amply demonstrates the achievements of the IAAP 'router' programme over the past two decades. With a truly Jungian commitment to mutuality and bi-directional learning, trainers and trainees alike record the fruits of this pioneering venture into the world of the other. Anyone interested in the relation between analysis and culture, and the rewards of cross-cultural dialogue, will greatly enjoy reading this book.

Warren Colman is a supervising analyst at the Society of Analytical Psychology and Consulting, and Consulting Editor for the *Journal of Analytical Psychology*

Here is a fascinating collection of reflections about the mutual influences of western psychoanalysts who teach colleagues in the East. Lots of experiences, questions, doubts, and exciting experiences from both sides are described and discussed. It provides basic texts for discussing the limitations but also the excellent prospects, together with the challenges of transformation for both in a cross-cultural partnership in teaching and understanding Jungian psychology. This book will surely open a broad discussion around these issues.

Verena Kast is Professor of Psychology at the University of Zürich, a training analyst at the C. G. Jung Institute, and a psychotherapist in private practice. She lectures throughout the world and is the author of numerous books, including *A Time to Mourn* and *The Nature of Loving*.

This outstanding compilation of articles gives voice to participants, including those of analysts, trainees, supervisors, and teachers, in the innovative and challenging adventure of providing training in Jungian analysis through Developing Groups in different cultural settings around the world. Neither over-idealizing nor shying away from often well-founded criticism, this book provides a balanced and nuanced overview creating an overall mosaic highlighting both the challenges and rewards of analytic work in different cultural settings on a theoretical, professional, linguistic, and personal level. Recognition of and appreciation for the bi-directional influence as a result of this foray into uncharted waters confirms that, despite very real limitations, psyche nevertheless finds creative, purposeful solutions and, that with flexibility, courage, and openness, the Jungian community has as much to gain and to learn from this experience as we have to offer.

Jung often stated that when he entered an analytic session, he tried to forget everything he had learned in order to be totally present and open to the individual before him. The very real challenge this represents is brought to life in the thought-provoking and deeply moving personal contributions of each author. This book is an absolute must for anyone with the willingness and courage to explore and question our ethnocentric perspectives and assumptions about analysis. I recommend it highly without reservation.

Tom Kelly is the Former President of the IAAP. He is a senior training analyst and supervisor, and past president of the Inter-Regional Society of Jungian Analysts and of the Council of North American Societies of Jungian Analysts.

Jungian Analysts Working Across Cultures

Jungian Analysts Working Across Cultures
From Tradition to Innovation

Edited by
Catherine Crowther
and Jan Wiener

LONDON AND NEW YORK

First published 2015 by Spring Journal Books

Published 2021 by Routledge
2 Park Square, Milton Park, Abingdon, Oxon OX14 4RN
52 Vanderbilt Avenue, New York, NY 10017

Routledge is an imprint of the Taylor & Francis Group, an informa business

© 2021 Catherine Crowther and Jan Wiener

The right of Catherine Crowther and Jan Wiener to be identified as author of this work has been asserted by him in accordance with sections 77 and 78 of the Copyright, Designs and Patents Act 1988.

All rights reserved. No part of this book may be reprinted or reproduced or utilised in any form or by any electronic, mechanical, or other means, now known or hereafter invented, including photocopying and recording, or in any information storage or retrieval system, without permission in writing from the publishers.

Trademark notice: Product or corporate names may be trademarks or registered trademarks, and are used only for identification and explanation without intent to infringe.

British Library Cataloguing-in-Publication Data
A catalogue record for this book is available from the British Library

Library of Congress Cataloging-in-Publication Data
A catalog record has been requested for this book

ISBN: 978-1-032-04981-6 (hbk)
ISBN: 978-1-032-04980-9 (pbk)
ISBN: 978-1-003-19545-0 (ebk)

Cover image:

The Clipsham Baboon by Jan Goodkin.
A representation of Thoth, the ancient Egyptian god of knowledge and writing. Carved in Clipsham stone, 2009.
www.jangoodkin.co.uk

Cover design, typography, and layout:
Northern Graphic Design & Publishing
info@ncarto.com

Table of Contents

Contributors xiii
Acknowledgements xix
Introduction
    Catherine Crowther and Jan Wiener, London, UK xxi

Chapter 1
Archetypes across Cultural Divides
    Murray Stein, Zürich, Switzerland 1

Chapter 2
Reflections on the Bi-Directionality of Influence in Analytical Work across Cultures
    Joe Cambray, Providence, Rhode Island, USA 19

Chapter 3
The Self and Individuation: Universal and Particular
    Ann Shearer, London, UK 37

Chapter 4
Pioneers or Colonialism?
    Henry Abramovitch, Jerusalem, Israel 53

Chapter 5
Cultural Complexes and Working Partnerships
    Catherine Kaplinsky, London, UK 71

Chapter 6
Understanding Group and Organisational Dynamics in Cultural Partnerships
    Stephan Alder, Berlin, Germany 91

Chapter 7
Issues of Cultural Identity and Authorship when Receiving Training from Other Cultures
    Gražina Gudaitė, Vilnius, Lithuania 111

Chapter 8
Women and Professional Identity in Russia
The Doll of Vassilissa the Beautiful as a Symbol of the Handover of Feminine Knowledge from Mother to Daughter
Elena Volodina, Moscow, Russia 127
Individuation as Container: The Search for Professional Identity in Russia
Natalia Alexandrova, Moscow, Russia 134

Chapter 9
Influenced, Changed, or Transformed? Reflections on Moments of Meeting in a Borderland
Tomasz J. Jasiński, Warsaw and Malgorzata Kalinowska, Katowice, Poland 143

Chapter 10
Bridging Two Realities: A Foreign Language in an Analytic Space
Kamala Melik-Akhnazarova, Moscow, Russia 161

Chapter 11
The Delivery of Training: Personal Experiences as a Trainer in Other Cultures
Angela M. Connolly, Rome, Italy 179

Chapter 12
Shuttle Analysis across Cultures
Personal Analysis by Shuttle: Can It Work?
Christopher Perry, London, UK 199
The "Frontier-Line in Human Closeness"
Ann Foden, London, UK 209

Chapter 13
An East-West *Coniunctio*: The Relational Field in Cross-Cultural Analysis
Liza J. Ravitz, San Francisco, USA 219

CHAPTER 14
GIVING VOICE TO PSYCHIC PAIN: THE BRITISH-MEXICAN
    CONNECTION, ON THE VICISSITUDES OF CREATING
    A HOME FOR STREET CHILDREN
    ALESSANDRA CAVALLI, LONDON, UK 237

CHAPTER 15
RETURNING TO CHINA
    JOHN BEEBE, SAN FRANCISCO, USA 255

CHAPTER 16
FROM TRADITION TO INNOVATION: WHAT HAVE WE LEARNED?
    CATHERINE CROWTHER AND JAN WIENER, LONDON, UK 273

INDEX 297

Contributors

Henry Abramovitch, Founding President and current Training Analyst in the Israel Institute of Jungian Psychology, is a professor at Tel Aviv University Medical School. He has served on Ethics and Program Committees of the International Association for Analytical Psychology (IAAP) as well as being President of Israel Anthropological Association and co-facilitator of Interfaith Encounter Group. He supervises routers in Poland and Moscow. His most recent books are *Brothers and Sisters: Myth and Reality* and *The Performance of Analysis* (Russian). abramh@post.tau.ac.il

Stephan Alder, Dr. med., Dipl. med., is a specialist in neurology and psychiatry, psychotherapy, psychoanalysis, and group analysis (German Society for Analytical Psychology, DGAP-IAAP; German Society for Group Analysis and Group Psychotherapy, D3G; and German Society for Psychoanalysis, Psychotherapy, Psychosomatics, and Depth Psychology, DGPT) with a private practice in Potsdam, Germany since 1991. He is a training analyst and lecturer for analytical groups (D3G), the C. G. Jung Institute Berlin, and the Berlin Institute for Group Analysis (BIG). He was a Balint group leader at the German Balint Society, 2008–2013, and a member of the board of DGAP and its president, 2014–2015. He is active in the analytical treatment of schizoid patients and patients with PTSD, with special focus on people suffering from psychosis and trauma. st-alder@t-online.de

Natalia Alexandrova, Ph.D., is an individual member of IAAP (2007) and a training analyst and supervisor for the Russian Society for Analytical Psychology (RSAP) and for the Moscow Association of Analytical Psychology (MAAP). She is co-author with Yuri Alexandrov of *Subjective Experience, Culture and Social Representation* (2009). Formerly a senior research fellow in the Institute of Psychology, Russian Academy of Science, she is now in private practice in Moscow. Her current interests include transgenerational trauma, identity and cultural identity, early development, infant observation, cultural complexes, and cross-cultural research. nataliya-alexandrova@yandex.ru

John Beebe, a past president of the C. G. Jung Institute of San Francisco, has taught about analytical psychology around the world, including China where he is the IAAP's Liaison to the Shanghai Developing Group. The author of *Integrity in Depth,* John was founding editor of the *San Francisco Jung Institute Library Journal* (now *Jung Journal: Culture & Psyche*) and the first American co-editor of the *Journal of Analytical Psychology.* He is co-author, with Virginia Apperson, of *The Presence of the Feminine in Film.* johnbeebe@msn.com

Joseph Cambray, Ph.D., is a past president of the IAAP; currently, he is the Regional Organizer for the IAAP in Asia. Dr. Cambray is on the editorial boards of *The Journal of Analytical Psychology*, for which he was U.S. Editor, and *The Jung Journal: Culture and Psyche*. He has been a faculty member at Harvard Medical School in the Department of Psychiatry at Massachusetts General Hospital, Center for Psychoanalytic Studies and an adjunct faculty at Pacific Graduate Institute. His books include his Fay Lectures, *Synchronicity: Nature and Psyche in an Interconnected Universe*, and a volume edited with Linda Carter, *Analytical Psychology: Contemporary Perspectives in Jungian Psychology*. cambrayj@earthlink.net

Alessandra Cavalli, Ph.D., trained as a child and adult analyst at the Society of Analytical Psychology. She worked in private practice at the Tavistock Clinic and the British Psychotherapy Foundation (BPF) in London. She published extensively in the *Journal of Analytical Psychology*, and was co-editor with Lucinda Hawkins and Martha Stevens of *Transformation*, published by Karnac in 2013.

Angela M. Connolly was an analyst of the Italian Society for Analytical Psychology (CIPA) with training and supervisory functions. She lived and worked in Russia from 1996–2001 and continued to teach, lecture, and supervise internationally. She was deputy editor of the *Journal of Analytical Psychology* and was on the advisory board. She was Honorary Secretary and Vice-President of the IAAP. She has published widely in English and Italian and has been translated into Russian and German. Her most recent works included "Cognitive Aesthetics of Alchem-ical Imagery" and "The Cognitive Aesthetics of Dream Meta-phors" published in *Understanding Jung* edited by Angiola Iapoce as well as "Out of the Body: Embodiment and Its Vicissitudes."

Catherine Crowther is a training analyst for the Society of Analytical Psychology (SAP) and past Chair of the Training Committee in private practice in London. She was a joint coordinator of the Russian Revival Project, training IAAP routers in Russia and was a joint IAAP liaison for the Developing Group in St. Petersburg. She is involved in supervision and teaching in the UK and in Eastern Europe. She has contributed articles to the *Journal of Analytical Psychology* and chapters to various books on analytical psychology. c.s.crowther@talk21.com

Ann Foden is a training analyst and supervisor with the British Jungian Analytic Association (BJAA) which is part of the British Psychotherapy Foundation (BPF). She is also a training analyst for the Tavistock Centre for Couple Relationships, the Association of Child Psychotherapists, and the Westminster Pastoral Foundation. She has a full-time private analytic practice in London working with individuals and couples. annfoden30@gmail.com

Gražina Gudaitė, Ph.D., is Professor of Psychology at Vilnius University and a Jungian psychoanalyst as well as President of the Lithuanian Association for Analytical Psychology. She is the author of several books and articles in analytical psychology, and co-editor with Murray Stein of *Confronting Cultural Trauma: Jungian Approaches to Understanding and Healing*. She has a private practice in Vilnius and coordinates and teaches in the Router Program in Lithuania. g.gudait@gmail.com

Tomasz J. Jasiński is an IAAP Individual Member and current president of the Polish Association for Jungian Analysis. He is a Jungian analyst in private practice in Warsaw, Poland. He is interested in non-verbal aspects of analytical processes, post-traumatic constellations in individuals and in the collective, and frontiers with spirituality in psychological healing. Currently, he is also a member of the Polish Association for Jungian analysis (PAJA) Training Committee and a teacher for the analytic training. jasinski@raven.edu.pl

Malgorzata Kalinowska is a Jungian analyst (IAAP Individual Member) and a founding member of the Polish Association for Jungian Analysis, where she holds the positions of Training Committee Member and Training Coordinator. She teaches for the PAJA training and works in private practice in Czeladź, Poland. She has worked in both clinical indoor and outdoor settings. She studies the relationship between trauma and culture and has published several articles on the subject in the context of the history of Central Europe. She is also interested in transcultural dimensions of analytical training and translates books on analytical psychology into Polish. kalinowska.gosia@gmail.com

Catherine Kaplinsky is a training analyst with the Society of Analytical Psychology in private practice. She has worked in various settings in the National Health Service (NHS), in a therapeutic community, in Child and Adolescent Psychiatry, and in a university. She has been practicing for twenty-five years and was a supervisor for the Russian Revival Project. She was born in India, brought up in South Africa, and immigrated to England in 1968. cathy@kaplinsky.com

Kamala Melik-Akhnazarova is a Jungian analyst currently working in private practice in Moscow, Russia, a member of the Russian Society for Analytical Psychology. She graduated from the philological faculty of Moscow State University in 1991. She combined the work of interpreter and translator with a long-term interest in astrology until these two paths finally brought her to the decision to be trained as a Jungian analyst. kamala27@rambler.ru

Christopher Perry is a Training Analyst and Supervisor of the Society of Analytical Psychology, of which he was also the former Director of Training. He is the author of *Listen to the Voice Within: A Jungian Approach to Pastoral Care* and several articles on analytical psychology and group analysis. He is interested in the interface between psychotherapy and spirituality. He lives and works in London. cjperry@orforduk.com

Liza J. Ravitz, Ph.D., is a certified Jungian child and adult psychoanalyst and clinical psychologist. She is a member of the C. G. Jung Institute in San Francisco where she teaches in the Analyst Training and the Continuing Education programs. Liza has clinical practices in San Francisco and Petaluma, California, where she works with children and adults. She is an Associate Professor at Sonoma State University and a teaching member of the International Society of Sandplay Therapists (ISST). Liza lived in Taiwan from 2012–2014 where she practiced Jungian analysis and taught as Visiting Scholar in Residence at Shieh-Chien University. dr.ravitz@gmail.com

Ann Shearer is a senior member of the Independent Group of Analytical Psychologists (IGAP) and lives in London. She has been teaching Jungian psychology in the UK and elsewhere in a variety of settings over many years; for four of these she was a visiting supervisor on the IAAP training in St. Petersburg. She is particularly interested in the ways in which mythological themes can contribute to psychological understanding. Her books include *Athene: Image and Energy* and (with Pam Donleavy) *From Ancient Myth to Modern Healing: Themis, Goddess of Heart-Soul, Justice and Reconciliation*. ann.shearer@zen.co.uk

Murray Stein, Ph.D., was President of the International Association for Analytical Psychology from 2001–2004 and President of the International School of Analytical Psychology in Zürich from 2008–2012. He is the author of *Minding the Self* and of many other books and articles on analytical psychology and Jungian psychoanalysis. He lives in Switzerland and is a training and supervising analyst with the International School of Analytical Psychology (ISAP) Zürich. murraywstein@gmail.com

Elena Volodina is a member of the Russian Society for Analytical and a Jungian analyst in private practice in Moscow. She is Senior Lecturer at the Department of Analytical Psychology at the State Academic University of Humanitarian Sciences,

Moscow. She is a shuttle analyst for routers in the Bulgarian Developing Group in Sofia. elenavolodina@yandex.ru

Jan Wiener is a training analyst and training supervisor for the SAP and BJAA in London. She is former Director of Training at the SAP. She was Vice President of the IAAP from 2010–2013 during which time she was Co-Chair of the IAAP Education Committee with special responsibility for Eastern Europe. She is author of numerous papers on subjects such as transference, training, ethics, and supervision, and her latest book, *The Therapeutic Relationship: Transference, Countertransference and the Making of Meaning*, was published by Texas A&M University Press in 2009. jan@janwiener.co.uk

Acknowledgements

For the idea for this book we would first of all like to offer our warm thanks to Murray Stein who has always supported all of our projects in Russia, has served as an external examiner for routers in Russia, and who took the initiative to suggest to Nancy Cater of Spring Journal, Inc. to publish a book based on our work in Russia. To our friends and Jungian colleagues in St. Petersburg and Moscow with whom and from whom we have learned so much, we dedicate this book. Without them, it simply would not have happened. We are proud of their achievements and will value our relationships with them for the rest of our lives. To colleagues in Krasnodar, Kemerovo, Ukraine, Poland, Serbia, Estonia, Kazakhstan, Romania and Bulgaria, we say thank you for inviting one or both of us to teach, to examine, or to work with them when their groups were in difficulty. We are grateful to the artists Jan Goodkin and Tim Harrisson for allowing us to use their sculptures as images of tradition and innovation on the cover of the book, and to Freddie Brazil and Jon Stone for their photography. We are also indebted to the meticulous editing work of Cat Westlake, Erica Mattingly and Alexis O'Brien. Finally, we thank all the authors who have worked with us creatively, energetically, and efficiently to produce this book which we hope does indeed pay tribute to both tradition and innovation.

—Catherine Crowther and Jan Wiener
February 2021

Introduction

JUNGIAN ANALYSTS WORKING ACROSS CULTURES

Catherine Crowther and Jan Wiener

The last thirty years have seen a burgeoning interest in Jungian psychology in different parts of the world where there have previously been few opportunities for serious study, personal analysis, or supervision for psychotherapists. Analytical psychology was unable to flourish in cultures with a history of political oppression or where totalitarian attitudes discouraged individuality—the cornerstone of our analytic tradition. In other countries, the lack of resources and educational opportunities or poor communications meant that there were few opportunities to learn or train. For many, leaving their countries of origin was not possible for political, geographical, or financial reasons, or because they did not speak languages other than their own.

One of the constitutional aims of the International Association for Analytical Psychology (IAAP) is to disseminate Jungian ideas as widely as possible. Spontaneous requests for training in Jungian psychology have prompted the IAAP to establish a number of Developing Groups (DGs) across the world—in the Chinese Region and Asia, Eastern Europe, Central Europe, South America, and North and South Africa. These projects are specifically designed to create viable structures, processes, and teaching programmes in which study and clinical training opportunities are available in new settings. They have come to be called "router training," a term coined by Murray Stein to permit individuals in regions where there are no established Training Societies to set off *en route* to become internationally recognised Jungian analysts.

Behind the vision of these programmes is a model of training where in the longer term, analysts could form new IAAP Societies when they are ready, which could in time take on training responsibilities within their own local culture and subsequently generate new Jungian life across the world. By 2015 this there were twenty-two Developing Groups and about 200 routers in training.

Given that the analytic tradition has been under pressure in the West during the past two decades, it has been encouraging to discover how much enthusiasm and curiosity there is now in other parts of the world not just for Jungian ideas, but also for hands-on learning from the clinical skills and experience of established Jungian analysts. In turn, Western analysts are not only excited at the prospect of visiting unfamiliar countries but are also discovering that these experiences contribute to development and change in their own theoretical ideas, teaching, and clinical practice back in their home societies. It is these mutual influences within new cultural partnerships, and the subsequent advances for theory and for cultural and clinical understanding emerging from them in the Jungian world, that are the central focus of this book.

It is our own deeply felt experience of working for thirty years in Russia that has inspired us to think and write about these issues and invite colleagues to contribute their own experiences and learning. Like the other authors in this book, we have been profoundly affected by the privilege of being introduced into another daily way of life and interacting with the professional attitudes and practices of a different culture. With all its unfamiliarity and challenges, we feel that our project of setting up and seeing through the IAAP training programme for routers in Russia has been one of the most rewarding, moving, and eye-opening experiences of our personal and professional lives. Our colleagues from the UK who participated with us in the training program (four of whom have contributed chapters to this book) all agree that through working in Russia our embedded beliefs, values, and practices have been subtly yet significantly altered. Although we went into the project with an enthusiastic optimism and trust that *something* would happen, we had no idea how enduring and far-reaching the outcome would be for everyone involved. This of course is reminiscent of Jung's analogy of the powerful effects of analysis on both the patient and the analyst as being like a chemical reaction of two substances where inevitably "both are transformed."[1]

Jung wrote,

> emotions are contagious because they are deeply rooted in the sympathetic system. ... Any process of an emotional kind immediately arouses a similar process in others. ... Even if the doctor is entirely detached from the emotional contents of the patient, the very fact that the patient has emotions has an effect on him. And it is a great mistake if the doctor thinks he can lift himself out of it. He cannot do more than become conscious of the fact that he is affected.[2]

Outreach experiences have revealed some of the limitations of our traditional models of learning in Western culture and are now resulting in the development of some unique, innovative, and tailor-made training programmes in other countries. Newer models and structures, where analysts venture out into uncharted territory taking Jungian attitudes and training into more isolated places, can lead to challenging but fascinating tensions between tried and tested frameworks for learning on the one hand and, on the other, some more inventive and adapted approaches in which the trainers have responded to the specific cultural values, social history, and individual wishes of professionals in each region. We hope that the reader will wish to learn more about these approaches that are evoked differently in each of the chapters.

As the title of our book suggests, we are interested to explore and evaluate the often delicate relationship between tradition and innovation through the voices of established Jungian analysts who have taught abroad, alongside the voices of those who have received their training within a cross-cultural partnership and have more recently become Jungian analysts. We invited contributions from Jungian analysts who trained in established institutes and now work in different parts of the world. All have a range and depth of experience of teaching, supervising, or offering personal analysis in other cultures. Other chapters come from those who came through a Developing Group or have trained within a Router Programme and who bring their personal experiences and their reflections, both positive and negative, of different aspects of their training and how these have affected their developing identity as

Jungian analysts. Each author was given a topic we asked them to cover, but how they wrote about it was up to them, resulting in a rich variety of chapters from analysts writing from different roles, in their individual styles, about the many different models and levels of training.

Yet what has emerged is a book not only about education but one that is even more alive to some of the broader shared problems of our day. Most of the cross-cultural projects around the world described in this book are in places where the population has in recent memory suffered huge political upheaval or social trauma. There is a lot to be learned regarding history, wars, different value systems, and social mores from each side about the other in the particular cultures in which the training partnerships developed. This book itself has become a reflective space and provides something of a social commentary on contemporary culture in a globalised world characterised not only by the possibility of increased cooperation, but also by conflict.

In our view, this is an important moment to step back and reflect collectively on the experiences of training and learning in different cultures. It is also a unique opportunity to explore the significance of how the different traditions in which we have lived and completed our own professional training influence our openness (or not) to consider new ways of learning and teaching and how easy (or not) it might be to adapt what we know in settings very different to our own. How can we best work towards cultural exchange and avoid culture clash? How readily can we understand the emotional needs of an unfamiliar patient population? What, concerning frame and process in our work and the relationship between the two, do we fight to retain and which of our precious beliefs must we lose because they are simply not really relevant in other cultures? Is a classical approach to analytical psychology more suited to teaching and learning in other cultures or is the developmental approach, underpinned by the centrality of the significance of early experiences, more useful? Can they be combined? How open-minded are we when encountering different attitudes to gender, sexuality, religion, ethnicity, family values, individuality, the collective? What are the strengths and limitations of offering "shuttle" personal analysis to routers in training within their region where sessions are irregular, sometimes via Skype, and sometimes with an interpreter in the room? What are the relative merits of group, as opposed to individual, supervision in countries where communism may have eroded notions of the individual's relationship to the group or the collective and when routers will need to find ways to work cooperatively in developing an organisational analytic culture in the future?

The first group to have completed the extended, multi-step transition from forming a Developing Group in analytical psychology to becoming an independent society of the IAAP with full training status is the Russian Society for Analytical Psychology (RSAP). Various other groups are at differing stages of this process at present. There remains within the Jungian world healthy debate and discussion about the value of cross-cultural educational projects, especially those where the teachers travel to the candidates, itself an innovation. We hope that the readers will evaluate for themselves the evidence either for or against this trend within the chapters of this book.

Here is a short summary of the topics in the chapters to come. It is impossible to do credit to the variety and depth of the writing. Many authors bring their deep scholarship of Jung's writings. They critically revisit what Jung derived from his understanding of other cultures, especially his absorption in Eastern philosophy, acknowledging his limitations and celebrating some creative mistakes. Murray Stein brings his knowledge of Jungian ideas to show how archetypal images of family and wholeness are interpreted differently in various cultures. Joe Cambray spans the world from his wide experience of different cultures and describes the dissemination of Jungian thinking as a process of bi-directional, reflective co-construction. Ann Shearer's chapter also draws on the different cultures she has worked in and explores her perception of the "familiar" universality of some human fundamentals, in tension with the "strangeness" of each new culture. John Beebe's deep study of the *I Ching* and Chinese philosophy allows him to muse on what Jung absorbed from the East and whether analytical psychology can return to China some of its own spiritual values after the Cultural Revolution. He marvels at the capacity of Chinese pictographs to condense complex metaphorical meanings, unlike the ego-driven Western languages. Several authors emphasise the unforeseen effects of trying to exchange ideas in a different language—both inhibiting and enlightening—and Kamala Melik-Akhnazarova brings her knowledge of several languages to write about the nuances of translation, both as an interpreter and as an analytic patient using a different language.

Several authors draw attention to their own multi-cultural origins and how their parentage, inheritance, or migration have coloured their attitudes to different cultures in which they have worked. Catherine Kaplinsky draws on her extensive knowledge about the concept of cultural complexes to show how different individual and societal histories play into both personal and cultural complexes. Henry Abramovitch brings his anthropological knowledge to bear, as well as his respect for storytelling and folklore, to explore perceptions of Jungian teachers as either pioneers or colonisers at different times in different societies. Stephan Alder, a group and Jungian analyst, approaches his inter-cultural encounters from the perspective of group and organisational theory as well as from under the shadow of his own East German origins. He writes personally of his awareness of the long-lasting effects on subsequent generations of the Second World War and the Cold War.

The ravages of the European conflicts of the twentieth century are also mentioned somberly by all those authors whose experiences come from the "other" side of the Iron Curtain. An intergenerational collective trauma continues to affect individuals and societies in "post-Soviet space," touching people's inner and outer lives. The Russian analysts Natalia Alexandrova and Elena Volodina recognise the long-term difficulties for women recovering from the coldly functional construction of the feminine and motherhood in the Soviet state. This is echoed by Gražina Gudaitė from Lithuania who reflects on living through the transition period from Soviet times. Jungian ideas helped to give gradual shape to a recovery from the previous Soviet devaluation of the uniqueness of each individual. Tomasz Jasiński and Malgorzata Kalinowska write of Poland as a precarious borderland, literally and metaphorically. They present a thoughtful critique of inter-cultural training models in a country where the enduring wounds of social and political history and cultural complexes are hard to contain.

Angela Connolly describes how her experience of training in Russia and Asia has had an impact on her clinical sensitivities and theoretical framework concerning personal identity, especially relating to the concept of the self in balance with the collective. Both she and Liza Ravitz add valuable additional insights from having lived and worked as personal analysts for a prolonged period of time in a different culture, Russia and Taiwan respectively. Being able to provide regular analytic sessions in both settings is in sharp contrast to the experiences of Ann Foden and Christopher Perry, "shuttle" analysts from London who give vivid accounts of how they and their patients weathered the privations of frequent long separations. They appraise both the shortcomings and possibilities of this way of analysis. Alessandra Cavalli brings her knowledge of infant and child development to the aid of professionals in Mexico whose psychological work with homeless street children is regularly supervised via email from London.

In the final chapter, we offer our own contribution from teaching and supervising in different cultures, especially in Russia, for more than twenty years. Because the region of our own involvement has looked east from London, we have not done justice in this book to the substantial development of Jungian work in South America or North Africa, and we regret these and other absences. We could not cover every Developing Group and Router Project in the world. Gathering up some of the similarities and differences in the themes emerging throughout the book, we bring an overall evaluation of this kind of work, its achievements, its limitations, and the implications for the establishment of cross-cultural partnerships in the future and its effects on our practice—often in a multi-cultural society—in our home countries.

An article by Philip Ball in *The Guardian* newspaper about the search for life forms in outer space provided an apt evocation of the work of all the authors of this book, in their explorations of those rewarding seams of our experience and learning that look to combine tradition and innovation.

> The occasional excitement at the discovery of seemingly 'Earth-like' exoplanets is understandable, but the fact is we have no idea what we're looking for, unless it is for ourselves.
>
> Not only is there no rigorous, scientifically agreed definition of 'life,' but there probably never can be. Like 'music' or 'culture,' the meaning depends on its fuzzy boundaries: try to pin it down and it evaporates. Yet this very lack of definition is ultimately one of the best reasons to be excited about the extraordinary diversity of worlds that our telescopes are revealing: above all else, they will force us to broaden our imagination about what life can be, and how we might recognise it.[3]

Notes

1. C. G. Jung, "The Psychology of the Transference" (1946/1954), in *The Collected Works of C. G. Jung*, vol. 16, ed. and trans. Gerhard Adler and R. F. C. Hull (Princeton, NJ: Princeton University Press, 1946), § 358.

2. C. G. Jung, "Lecture V, Tavistock Lectures" (1935), in *The Collected Works of C. G. Jung*, vol. 18, ed. and trans. Gerhard Adler and R. F. C. Hull (Princeton, NJ: Princeton University Press, 1946), §§ 318–19.

3. Philip Ball, "The search for life beyond Earth is on, but are we ready for what we'll find?," *The Guardian*, Aug. 8, 2014.

CHAPTER 1

ARCHETYPES ACROSS CULTURAL DIVIDES

MURRAY STEIN

While it's very exciting and enticing to travel to distant lands to teach the theory and practice of analytical psychology, it's also an eye-opener. Psychoanalysis and analytical psychology are products of modernity in the West. Their discoveries were made and their methods of treatment were forged under the specific cultural conditions of middle Europe during the late nineteenth and early twentieth centuries. Sitting in our offices and classrooms in the West today one hundred years later, we may be oblivious to the reality that our modern and secular cultural values and attitudes have not been taken up and adopted equally everywhere in the world. Even as globalization has carried in its train many of our habits of thought pertaining to technology and science and has planted them in widely dispersed regions of the world, the psychological and social assumptions characteristic of Western modernity have not necessarily traveled along with them. For many people in other than Western nations, psychoanalysis and psychotherapy look exotic and foreign, even threatening to basic assumptions, especially religious and traditional ones based in ancient habits of living and thinking. To what extent does this inhibit the possible transfer of our analytic methods and psychological perspective to other parts of the global stage in the twenty-first century?

In our recent efforts to teach analytical psychology throughout the world, we have come into contact with many peoples who are not so deeply influenced or impressed by our own local philosophies and social understandings. In many cases, we (I am referring here to efforts on the part of members of the International Association for Analytical Psychology) have encountered significant differences in mentality that have often provoked deep resistance, whether vociferously voiced or signaled by silent withdrawal, to our methods and ways of conducting Jungian psychoanalysis. As we work with people from cultures really different from our own, whether in supervision, analysis, or clinical seminars, it has become starkly obvious to us that at significant points, we do not share similar assumptions about many things that belong specifically to our ways of thinking and working. While we may start out with the fantasy of teaching, we end up with the need for dialogue.

At first we might be shocked and surprised because our theoretical and clinical assumptions seem so self-evident and obvious to us. But the opposite is true. There is nothing self-evident about them at all. They are not universals. They are built on earlier convictions, beliefs (religious and philosophical), and cultural habits that grew up in the West over centuries. It is within and as part of these established cultural patterns that we have been trained to work as analysts and therapists. Psychotherapy must be conducted in a particular way, we are inclined to believe, or it will not be transmutative and nothing will change or develop in the client's inner world. For us, these rules of practice may seem transcribed like the Ten Commandments in stone, but for others who are foreign to our traditions they may appear socially unacceptable or even bizarre. To question these assumptions puts us on a line of thinking that must lead to profound considerations about the fundamental nature of the human psyche and its possibilities for transformation in the general human population. Teaching in cultures widely different from our own, we are challenged to question and perhaps even forced to relativize our convictions, at least to a degree, and to rethink many of our preconceptions and often un-reflected upon conventions of practice. Old habits die hard on both sides—that of the student and that of the teacher. In the end, if we are fortunate, we may arrive by a kind of alchemy at an agreeable compound of the positions.

Why do these cultural differences exist at all? Where do they come from and how do they arise and become embedded in social groups? This question has many answers depending on one's preferred modes of understanding the origins of human culture and its varieties of expression. From a Jungian perspective, we hypothesize that all human forms of culture are manifestations of underlying universal archetypal patterns of human behavior, cognition, and imagination. Partly they are creations of human beings who have needed to adapt to different specific environments, and partly they are manifestations of underlying possibilities that have emerged from the universal (archetypal) psychic matrix. They are a product of the interaction of psyche with object world. All cultural forms of behavior and attitude reflect psychic constants, but there are innumerable variations on these embedded themes because of environmental differences and historical processes in given social groups.

In this chapter, I will speak of some of the problems that arise in the supervision, training, and analysis of persons from cultures different from our own in the West, and I will do so in the light of some archetypal constants that might play a role in bridging differences and solving some of the dilemmas we confront as we seek to take analytical psychology into the global community.

Bridging cultural differences is made somewhat easier by recent developments in Western attitudes and mentality. Today, rapid change in cultural preferences and sharp differences of opinion with regard to human and political values are becoming the norm in the West. Europe and North America and their direct descendants (which is what I mean primarily when I refer to the West) have become multicultural to an unprecedented degree in the last few decades, and this has brought in its wake an urgent need to relativize all religious and traditional cultural values and to find a way of containing traditional cultural differences within a larger, embracing, mandala-like cultural attitude.

On the surface of rational consciousness, this makes sense. But it is a different picture at deeper levels of unconscious assumptions and acculturations. We may believe that our standards of practice are of only relative validity, but they are deeply embedded in us and remain unreflected upon until they are challenged. To some extent, however, we have grown accustomed to living in cultural flux and with neighbors who are different from ourselves. We see this as individuation, a step forward in the evolution toward wholeness, on a cultural level. We realize that cultural standards and values are something that passed for absolute truth in previous centuries in the West, but do so no longer.

For example, in the time when colonialism was in full swing, the norms of the imperial nations were unquestioned and were imposed on subjected peoples without a second thought about their universal validity for everyone. The colonial powers believed without serious question that they were raising the cultural standards and attitudes of the colonized peoples by educating them to a higher level and even saving their heathen souls by baptism into the Christian faith. Today, we totally reject this view of the civilizing influence of the colonial powers; if anything, we demonize this position for totally overlooking the values and religious beliefs of the cultures they were colonizing.

Jung affirmed the spiritual and psychological value of cultures very different from our own in the West.[1] They are, he argued, not only equal to ours on a relativistic measure—i.e., made up of habits that are, like ours, without ontological standing—they actually offer something that we lack. In some respects, they are to be deemed superior. They may remain in much closer contact with archetypal (i.e., religious or mythological) ideas and images that give them great power and grounding in the fundaments of the human psyche. In a sense, Jung reversed the colonialist position and instead of placing Western culture at the top, he moved it to a much less exalted position.[2] He brought the West down to a much more humble place and forced us to rethink our attitudes vis-à-vis, for instance, the religions of the East. Spiritually speaking, we are poverty-stricken and they are wealthy. The West, in Jung's view, has severed connections with nature, which "primitive cultures" have not;[3] the West has lost contact with the archetypal and the spiritual, which the East has not. What the West has developed is rational, positivistic science, and it has purchased this at the expense of deep psychological awareness of body and soul. When other traditional cultures look at the West, they admire the technology but shudder at the spiritual poverty.

At the same time, Jung affirmed a common psychic ground for all of humanity in what he chose to refer to as a "collective unconscious."[4] This was not a return to absolutes as these had been conceived and believed in during earlier phases of Western history, in the Middle Ages for example, but rather it was a perception of underlying (archetypal) patterns and motifs that play beneath the surface of all cultural traditions and religions, a common source of inspiration and motivation. Inhering within this psychological foundation of humanity are archetypal images that tie us all together, such as Mother, Father, Family, Hero, the Couple, Good and Evil, the Center, and Wholeness. Each of these archetypal patterns finds expression in some manner in every culture created by our species. It is therefore to this level that we refer when we express the sentiment that all men and women are brothers and sisters, that we all belong to humanity equally and share a common psychological heritage. It is on this basis that we can offer psychotherapeutic assistance to people of all cultures. Our psychotherapy must be cognizant of cultural difference but grounded in underlying commonality, otherwise we will fall either into the trap of cultural imperialism or find ourselves totally irrelevant. This is a Scylla and Charybdis problem. Awareness of archetypal patterns may help us bridge cultural divides.[5]

It can be a great challenge to work in analysis with deep cultural differences between analyst and patient. Frequently, analysts must confront their own particular cultural limits and decide whether they can continue working with a person who exceeds what they feel is tolerable. I once supervised a candidate in training who was working with a middle-aged woman of European and Christian descent, much like herself, but who had converted to Islam and was married to a much older man from a Middle Eastern country. The therapist was stumped by the patient's attitude toward what we would generally consider to be the husband's quite severe verbal (but not generally physical) abuse. The patient was a highly accomplished academic on the one hand, and yet she behaved like a submissive dependent child toward her senior husband on the other. In supervision we worked on the assumption that a pathological masochistic complex rooted in her childhood relationship with her father was operating to distort the patient's thinking and feeling, leading her to split off and dissociate from her received abuse in order to maintain her relationship and protect her self-esteem. Normally, this could have been worked on, and possibly resolved, in the analytic process, if the patient and analyst would have shared a perspective on what constitutes intolerable behavior. While under the influence of the husband's cultural standards and expectations, however, she took the abuse as a normal state of affairs. In the other sectors of her life, such behavior would have been intolerable. In the cultural context of the marriage, moreover, divorce was not an option. Her choice was to travel as much as possible for her career and live in bondage when at home. I came to regard this as an irresolvable impasse in the analytical process because it could not be challenged effectively. Individuation possibilities seemed to have come to a halt in the face of a cultural pattern that does not regard women as fully human and capable of mature wholeness; however, in the course of time my student and I both realized that a much more profound religious need was being addressed by this woman's seemingly irrational adherence to cultural patterns so different from the ones we are familiar with and consider norms. She was in fact sacrificing her own inherited cultural identity and freedoms as a woman on the altar of what she experienced as a transcendent value of spirituality in a tradition quite foreign to her own native one. We could not force our convictions on her without damaging her (to us seemingly bizarre) individuation process.

I have never forgotten this case. As Jung taught us, the human psyche is a great mystery and has its own ways. We must be humble in the face of such paradoxes and try to understand them from a deeper perspective that steps beyond our personal cultural assumptions and preferences. Fortunately, this particular candidate, a mature woman with much experience as a psychotherapist, was able to continue working with this patient on the Jungian assumption that the self is at work even when we do not understand it.

Mother and Father as Archetypal and Cultural Images

I will address now some of the archetypal patterns underlying all cultures and reflect on how differences from culture to culture in their manifestations and formative influences have an impact on how teaching and training psychoanalysis and psychotherapy is carried out or problematized. The first of these is Mother and Father, the Parents.

Every culture in the world knows of mothers and fathers. Beyond their biological significance, they are psychic images and forces. Jung and those following his line of thought speak of the mother archetype and the father archetype, which have mythological expressions in many cultures. Jung wrote of the mother in this sense in his essay, "Psychological Aspects of the Mother Archetype,"[6] and Erich Neumann authored the standard work on this image in *The Great Mother*.[7] Every culture has a more or less generally accepted image of motherhood, whether explicit or implicit, and these seem to be quite similar across cultures. Fatherhood seems to be less defined and more variable from culture to culture, although Father Gods appear quite universally.

Mother and Father, however, generally show two aspects—one positive and the other negative. Typically, these contrasting features are represented in a culture's myths, fairy tales, art, and other imaginative expressions. Most traditional cultures create images of divine Mothers and Fathers and hold these up as ideal or fearsome models. Modern cultures depict the images of ideal or demonic mothers and fathers in movies, cartoons, or in other popular media. Generally, the definitions of what constitutes the maternal or the paternal are unquestioned in a culture; they are simply enacted and lived. We also know that while there is a mother and a father archetype in the background, the actual mothering and fathering that is done by individual women and men in different cultural contexts, while embedded in a sort of instinctual base that humans share with other mammals, is quite flexible and capable of a wide variety of expression.

Archetypally, Mother signifies containment (the womb and the lap) and nurturance (the breast) on the positive side, and restriction (smothering) and death (the tomb) on the other. Both images can be found in mythologies worldwide. Father, archetypally speaking, signifies the demand for achievement and performance, the rule of law, and hierarchical social structure with all of their concomitant features, positive and negative. Every culture has both maternal and paternal features, but the degree of dominance of one or the other varies considerably.

In our efforts to train psychotherapists in cultures different from our own, we need to keep in mind the general preferences and emphases the student's culture places on the mothering and fathering principles. Some cultures are more oriented than others toward the maternal than the paternal principle and toward the values that derive from maternal images. We can even speak of mother cultures, where the images of the Great Mother predominate and importantly contribute to the basic values of the culture as a whole. In these, adult women play a controlling role in the home and community, whether explicitly or behind the scenes. The collective psyche of these cultures draws heavily on the mother archetype for its forms of human interaction, including social and political life and policy. The State is constituted as a nurturing container for its inhabitants, and expectations of nurturance and support develop in the population accordingly.

In taking up the profession of psychotherapy, the people in these cultures tend toward strong maternal countertransference attitudes toward their patients. The patient is seen as an object of maternal nurturance and caretaking, and there is a minimization of legal and other external interference in this dyadic relationship. It is a tight couple with little room for a third within the sacred embrace of mother and child. The maternal therapist becomes protective of her clients like a mother bear of her cubs. It is a highly personal relationship. In this cultural setting, supervision can be seen as an attempt to apply general rules to what is after all a unique personal relationship. Outer authority is rejected or subtly defended against in favor of protecting the intimacy of the analytic couple. Here insistence on professional boundaries and of the frame of professional work tends to be a major challenge requiring rigorous efforts by supervisors to correct the tendencies to merge personalities that are inherent in the mother-infant dyad. The supervisor has to understand that maternal aspects of psychotherapy, like holding and nurturing and supportive encouragement, are important and based on archetypal and cultural foundations, but they are not all there is in the ideal analytic process. They need to be compensated from another angle. Cultural biases may inhibit the model of analysis as developed in the West, and compromises will need to be considered. These particularly have to do with frame and boundary issues—time, setting, payment of fees, interactions outside of therapy setting, etc. Supervision of these is generally seen as meddling in the sacrosanct precincts and to be discarded or resisted.

I recall trying to supervise a student from a Latin American country, a typically strongly mother-identified culture where the men pretend to be in charge with their gaucho posturing but are actually controlled by strong feminine (mother) figures in the background. The relationship between therapist and patient struck me at first as inappropriately intimate. Sessions would open with a hug; gifts were exchanged in sessions; time setting and maintaining were regarded as highly flexible; and payment of fees was casual and frequently ignored. When I tried to challenge these practices, I discovered to my astonishment that these were considered normal. In fact, I learned, if a therapist offered to shake hands rather than embrace a patient, this would be taken as an insulting rejection. The supervision of this student forced me to reflect on my North American practices because, as I discovered, the therapeutic outcomes were actually very positive. In the end, I did succeed in at least bringing some reflection into the practices so different from my own, and I could accept to a degree that frame issues can be different and yet serve to contain the analytic work. North American culture, founded by north European Protestants, is weighted much more toward patriarchal "father" rules than South America, where the dominant religious images are of the Great Mother. This plays out in political and social structures, and it of course also affects the analytic relationship.[8]

In cultures where the mother archetype is not so firmly installed and embedded in habitual interpersonal relationships, the atmosphere of the therapeutic setting is quite different. The Great Father dominates the archetypal background. One thinks of cultures strongly shaped by Northern European Protestantism, Judaism, or Confucianism, for example. Here cultural patterns of attitude and behavior are quite different from those in Mother cultures, as for example in the Mediterranean. The therapist's attitude will be more impersonal and professional, keeping back at a greater psychic distance in a well-defined persona as doctor or technical specialist. Supervision of students in this sort of cultural setting might be aimed at loosening them up emotionally and encouraging them to pay more attention to countertransference feelings and intimacy issues and needs, rather than focusing on the social adaptation and behavior of their clients. Here examination of countertransference feelings especially will be an area of pressing concern in supervision. Of course this will be resisted, at least for a time, since the cultural bias assures a large degree of discomfort with this direction. Maintaining a wall of silence is a typical means of defense.

In a private conversation with a professor of psychotherapy in an Asian country, I heard the following extraordinary statement: "We teach theory in the classroom, but when the students go into a room to conduct therapy sessions and close the door we have no idea what goes on." Students are apparently unwilling to risk embarrassment in supervision, and confessing private countertransference feelings and reactions especially is far too potentially shaming. Strongly patriarchal cultures emphasize principles of authority, law, emotional distance, and hierarchically structured relationships. There is little room for the intimate and personal in professional life; that is left for after hours.

In the consulting room, therefore, the atmosphere tends to be formal. There are expectations of behavioral change in the patient, and perhaps punishments for failure. In cultures such as this, supervision attempts to decrease emotional distance with patients and to foster an attitude of paying closer attention to emotional nuance in the therapeutic relationship. Training in countertransference awareness becomes of paramount importance. In our standard view in the West, psychological growth in analysis is greatly facilitated by the personal relationship between client and doctor. Again, cultural habits may place limits on how far supervision can affect the approach of the student therapist in a Father culture. Emotion is uncomfortable, especially when shared in a professional setting. In extreme cases, psychotherapy as we know it in the West would be impossible due to cultural inhibitions.

In our Jungian trainings conducted worldwide, we have confronted both types of culture. Each has its particular challenges, and in each we have had to strike compromises. What we would like to see is a balance between the maternal and the paternal principles in our analytical work; however, an analyst must, after all is said and done, work within a given cultural framework. Some Jungian analysts, especially those trained in Institutes in the West, have become strong critics of their own cultures' one-sidedness, but they have to be careful not to push too hard or they will be out of work.

INDIVIDUATION: AN ARCHETYPAL PROCESS OF PSYCHOLOGICAL DEVELOPMENT

One theme dear to the heart of most Jungian analysts is individuation, but this perspective on psychological development is certainly not without a problematic aspect when traveling across cultures. Far more than Mother or Father archetypal images and patterns, the archetypal process we know as individuation in Jungian psychology[9] is foreign to many cultures of the world. This is because it focuses on the individual and not the group or family. Individuation implies strong tendencies toward separating individuals psychologically from the social groups in which they are raised and nurtured.

Traditionally, human groups such as tribes at first and as homogeneous nations later have been tightly organized around strictly culturally defined collective values and practices embedded in ritual and religion. In these contexts, individuals are looked upon as group members rather than entities unto themselves; likewise, individuals look upon themselves as parts of a social network rather than as autonomous selves. Personal identity is defined by relationships to significant others such as parents, siblings, and ancestors. The instinctual/archetypal impulse to individuate is tightly regulated within strict cultural confinement and with severe limits and penalties for infringement. In the West, the spirit of individuation was released with the Renaissance, the Reformation, and the Enlightenment, and it has been growing in power and influence, for better or worse, in the last couple of centuries. This has been accompanied by the decline in the influence of religious and other cultural authorities that had previously kept a tight lid on this type of archetypal energy. Jung speaks of the dangers and positive values of this development in "The Spirit Mercurius."[10]

Jungian analysis offers a way of working with powerful individuation energies and impulses in a relatively closed and contained space, and it seeks to metabolize the shadow side of individuation (which, if it takes over, leads to excessive individualism) and link it to the project of creating a personal sense of wholeness. This involves a complex set of methods and techniques, such as dream interpretation and active imagination, which promote at first a sense of separation and even isolation of the individual from communities and collectively inspired identities before they lead to a deeper level of solidarity with humanity and its evolution toward greater consciousness. This initial stage of individuation can be perceived as highly threatening to cultural authorities as well as to internalized cultural images and their enforcers.

The awareness of personal individuation as a possibility for people of all ages and both sexes is new and threatening in traditional cultures. Whether a culture is dominated by the archetype of the Mother or the Father, the heroic energies of the emergent child breaking free of parental restraints and becoming fully an individual and independent of external authorities understandably provokes anxiety and defensive reactions. In supervision, an appreciative understanding and support of individuation processes active in the students' clients can be read as antisocial and even immoral. Here Jungian teachers have had to tread lightly. Homophobia, for example, can be easily aroused by open reflections on the individuation of gay and lesbian clients. Traditional cultures tend to pathologize individuation if it in any way steps out of line with their conventional norms, and at times even professional associations have been enlisted to enforce adverse or critical judgments against attitudes and behavior that run against the consensus. One thinks, for example, of Soviet psychiatry that was used by the State to quell political dissent: ideological divergence was considered to be a form of mental illness. Individuation in its fullest sense requires the freedom to become oneself in every respect, including emotional, spiritual, and intellectual dimensions.

Lest one point the finger at others without taking stock of one's own culture's biases and inhibitions, I recall that there was a movement in American psychiatry when the DSM-III was being formulated in the 1970s to designate introversion as a pathological condition. This of course represents a shadow enactment on the part of the American cultural consensus, which considers extroversion to be healthy and introversion sick. An attempt to "cure" introverts and make extroverts of them, however, will seriously stunt or damage a normal individuation process that runs its course according to the natural pattern of an introverted orientation. Naturally, one hopes for an eventual balance, more or less, between introversion and extraversion as an outcome of an attended and conscious individuation process, but at some stages of life this may not be suitable. After significant protest by Jungian analysts (among others), the suggestion to pathologize introversion was withdrawn by the creators of DSM-III.

Individuation has specific and general features. Each individual human being's life creates a unique rendition of the archetypal template that is lodged within the psychic matrix. Jung discusses the complexity of this process in many of his writings,[11] and other analytical psychologists since his time have carried the discussion further. It is clear by now that the centrality of the individuation process is a non-negotiable feature of Jungian thought and practice. For analytical psychologists, this is a universally valid feature of the healthy growth of personality, and it cannot be sacrificed for the sake of cultural adaptation. Teachers and supervisors, however, may mute some aspects of individuation for tactical purposes in some areas of the world, even as it remains implicit in their attitudes toward psychotherapy. In the long run, we believe, individuation will prevail as a central feature in the evolution of human consciousness.

Relationships: The Couple as Archetypal Image

Because human beings are for the most part basically social creatures, relationships are a key feature of a balanced and fulfilled life. The ways in which relationships, especially intimate ones, are constructed and lived, however, are remarkably variable. For example, monogamy is a fundamental feature of social life in Western cultures, but this does not mean it is archetypal and therefore universal. It is a social construction built up from the archetypal unit of the dyad—the couple. This is lived in many ways. Increasingly, in the countries of modernity and postmodernity, serial coupling has become an accepted practice. In some traditional cultures, polygamy is the norm, that is, a number of couplings form around a central male or female figure. Psychotherapists have learned to be quite tolerant of many variations on the theme of coupling without necessarily pathologizing them, although they will want to reflect on their qualities and motivations. In some cultures, the options of coupling are quite limited and restricted by religious and social customs. Here, supervision requires careful sensitivity to cultural norms in the place. The important thing for psychotherapeutic purposes is that coupling facilitates growth and individuation in the client, but this caveat can create major difficulties especially when it comes to what therapists would generally consider to be abusive behavior within the couple.

The Shadow

A question that inevitably comes up in supervision and training has to do with shadow awareness and integration. In Jungian psychology we are trained to observe the shadow in ourselves as therapists and in our clients. This refers to a particular type of sensitivity to the falseness of human pretensions and facades. It is a habitual self-critical attitude that is aimed at greater consciousness. Jung said of Freud that he was the master of shadow awareness and exposure.[12] The shadow is an archetypal feature of the human personality, which means it is universal and cannot be eliminated by social, cultural, or religious influences. Different cultures have different methods for dealing with it, however. Most religions are acutely aware of the problem of evil and invest great energy in attempting to keep it under control and at bay. None have succeeded to date. Nor has analysis. What careful consideration of the shadow in the analytical container can and should do, however, is to raise it to greater consciousness and thereby reduce the tendency toward splitting or blind acting-out of shadow impulses.[13] The technical problem that analysts face is bringing the shadow into consciousness without arousing violent defenses in the client, on the one hand, or creating devastating humiliation, on the other. M.-L. von Franz writes in this regard about practicing "bush manners" when approaching the inferior function,[14] which is tantamount to the shadow—that is, raising a white flag and giving plenty of warning and assurances that no harm or humiliation is intended. It is a narrow and precarious line to walk.

Supervision must walk the same delicate path. Students from cultures other than the supervisor's may unwittingly misunderstand a supervisor's motivations for pointing out their shortcoming and errors. Supervisors may receive a projection of arrogance or cultural imperialism, even though they may be only trying to correct what appears to be a faulty technique or approach to a client. Conversely, supervisors may project the same shadow features onto students who blithely refuse to consider their suggestions. In Latin American countries, for instance, where North Americans are widely regarded as imperialistic "gringos," this dynamic has happened with frequency between students and supervisors/teachers. The dynamics are ferocious. The supervisor's misunderstanding may be that the students are frightened or insecure, not arrogant. The cross-cultural learning process is fraught with possibilities for mutual shadow projection, and here knowledge of cultural attitudes and habits can be critically important. Reading one another accurately in the training situation when teacher and student come from different cultural backgrounds is challenging. Fantasies tend to drift to cultural stereotypes, and in time they may become solidified into unbreachable misperceptions. (Of course, this happens in training programs in Western cultures as well because of the diversity of cultural backgrounds among students.)

The Archetype of Wholeness—The Self

Ultimately, what we are reaching for in all our therapeutic endeavors is to build up and establish a conscious sense of psychological wholeness. This development too is based upon an archetypal foundation: the increasing awareness of the archetype of the self as the individuation process progresses. Erich Neumann speaks of this phase of individuation as "centroversion," which entails greater consciousness of what he calls the "ego-self axis."[15] A sense of wholeness means acceptance of one's complexity as a personality, including one's internal divisions, contradictions, and conflicts. This exceeds what is usually meant by identity, which is based on identification with other persons, cultural values and myths, religious figures, etc. Wholeness is universally represented by the circle, which Jung termed the mandala after Indian and Tibetan practice.[16] The mandala contains all that is, and holds it in a unified container. The sense of wholeness is the contrary of one-sidedness, and it is the cure for neurosis, which is based on non-acceptance of psychological complexity.[17]

Many cultures hold up one-sidedness as an ideal—the ideal of holiness, or unambiguous devotion to king or political party, or undivided patriotism to one's country (*My country, right or wrong!*). Working as supervisors and teachers in cultures where such one-sidedness is dominant and wholeness is seen as indecisive, weak, or even criminal, we must keep an eye on cultural sensitivities. In a sense, therapy is subversive and undermines cultural hegemony over individuals and their personal development. We walk on the margins, therefore, in our activities as exponents of the practice and theory of analytical psychology. I have found the Greek god, Hermes, to be a helpful mythological guide in these endeavors.[18]

Conclusion

Even while we as Jungian psychoanalysts, teachers, and supervisors work for individual self-development, we also work with the notion of changing cultures to the extent that they can participate more fully in the emergent global culture. Our efforts with small groups of professionals who have an interest in Jungian ideas and methods may help to infuse their cultures with values such as the freedom to pursue individuation for all. We are not arguing for political arrangements or social change per se, but we do want to advocate strongly for psychic space to realize as fully as possible the potentials of the self. We also realize that in carrying out our task to teach the theory and practice of analytical psychology worldwide, we are changed. The relationship between teachers and students must remain dialectical so that both can learn and grow from the experience.

NOTES

1. In this passage, I am summarizing Jung's comments scattered through many of his works, among them "Yoga and the West." C. G. Jung, *The Collected Works of C. G. Jung*, vol. 11, ed. and trans. Gerhard Adler and R. F. C. Hull (Princeton University Press, 1969), § 859.

2. This view is stated especially forcefully in Jung's *Berlin Seminar*, to be published soon in the Philemon Series. In *The Psychology of Kundalini Yoga* (Jung, Princeton University Press, 1996), Jung argues that the East has gone far beyond the West in its development of interiority and spirituality. The West can boast scientific and technological superiority, but the East has gone further spiritually and psychologically. In other words, the West has much to learn from Eastern philosophies, religions, and cultures with regard to inner development.

3. Jung, "Archaic Man," CW 10, § 104.

4. Jung, "The Archetypes and the Collective Unconscious," CW 9i.

5. Many Jungian authors have reflected on this issue. *The Cultural Complex*, edited by Thomas Singer and Samuel L. Kimbles (Brunner-Routledge, 2004) offers a representative sampling with eighteen articles by authors from a variety of backgrounds and Jungian "schools."

6. Jung, CW 9i, §§ 1, 48.

7. Erich Neumann, *The Great Mother* (Princeton, NJ: Princeton University Press, 1955).

8. For more on this topic, see Murray Stein, "On the Politics of Individuation in the Americas," in Thomas Singer and Samuel L. Kimbles, eds., *The Cultural Complex* (Hove and New York: Brunner-Routledge, 2004), pp. 262–73.

9. Murray Stein, *The Principle of Individuation* (Asheville, NC: Chiron Publications, 2006).

10. Jung, CW 13, § 239.

11. See especially Jung's essay, "Conscious, Unconscious, and Individuation," in CW 9i, § 489. There he offers a succinct definition: "I use the term 'individuation' to denote the process by which a person becomes a psychological 'in-dividual,' that is, a separate, indivisible unity or 'whole'" (§ 490). He then continues to explain the complexities of the process, which includes overcoming splits and divisions between ego-conscious and the unconscious.

12. Upon Freud's death in September 1939, just weeks after the outbreak of WWII, Jung wrote, "He aroused a wholesome mistrust in people and thereby sharpened their sense of real values. All that gush about man's innate goodness, which had addled so many brains after the dogma of original sin was no longer understood, was blown to the winds by Freud, and the little that remains will, let us hope, be driven out for good and all by the barbarism of the twentieth century. Freud was no prophet, but he is a prophetic figure. Like Nietzsche, he overthrew the gigantic idols of our day, and it remains to be seen whether our highest values are so real that their glitter is not extinguished in the Archerontin flood." (CW 15, § 69).

13. Erich Neumann writes about this passionately in his work, *Depth Psychology and a New Ethic* (New York: G. P. Putnam's Sons, 1969).

14. M. -L. von Franz, *Psychotherapy* (Boston and London: Shambhala, 1993), p. 125.

15. Erich Neumann, "The Psyche and the Transformation of the Reality Planes: A Metapsychological Essay," in *The Place of Creation* (Princeton, NJ: Princeton University Press, 1989), pp. 3–62.

16. Jung, "Concerning Mandala Symbolism," in CW 9i, §§ 627–712; and Jung, "Mandalas," in CW 9i, §§ 713–18.

17. Andrew Samuels, Bani Shorter, and Fred Plaut, *A Critical Dictionary of Jungian Analysis* (London and New York: Routledge & Kegan Paul, 1987), p. 99.

18. Murray Stein, *In MidLife* (Asheville, NC: Chiron Publications, 2015).

CHAPTER 2

Reflections on the Bi-Directionality of Influence in Analytical Work across Cultures

Joe Cambray

Introduction

As analytical psychology has moved out of its traditional locales, increasing challenges have emerged in working across cultures. Few analysts are formally trained in multicultural approaches, which often require years of post-graduate work in fields such as anthropology, sociology, cultural studies, as well as acquiring appropriate language skills, to start. Yet genuine, deep interest, need, and willingness to sacrifice much for the sake of being educated and trained in analytical psychology continue to be brought forward by bright, committed clinical professionals around the world. There has been a reciprocal desire from numerous analysts to respond in as effective a manner as possible given the limitations and constraints on all parties involved. This chapter will attempt to assess an underreported value from these encounters in the International Association for Analytical Psychology (IAAP) Developing Groups (DGs): the bi-directionality of influence emerging out of the cross-cultural engagements.

Diverse applications of the concept of bi-directionality are well known in many fields of research including psychological studies. For example, Georg Northoff in *Philosophy of the Brain* (2004) notes "The 'principle of bi-directionality' consists in the necessity of bi-directional linkage between philosophical theories and neuroscientific hypotheses and thus between logical and natural conditions."[1] Similarly, recent studies have enhanced our understanding of the bi-directional influences between the human microbiome, especially the bacteria of the human gut and the brain, with ramifications for understanding free will in human emotions and behavior.[2] For an analytically-oriented approach, one salient area derives from research on families, especially bi-directional influence observed between parents and children, and comparisons made in varying cultural contexts. Thus, Beebe and Lachmann's pioneering work on the co-construction of consciousness in mother-infant dyads helped to establish an experimental basis for their extension of these observations to adult psychotherapy.[3] Linda Carter has incorporated this perspective into a range of Jungian reflections on relationships, from countertransferential explorations,[4,5,6] to those between artists.[7] Tronick and Beeghly investigating meaning-making in infants have contextualized their findings with discussions of cultural contributions in different mother-infant dyads.[8] Tronick et. al., also examined various dyads, including psychotherapeutic partners, as forms of self-organizing systems, which can generate richer, more complex states of mind for both partners, a topic we will return to later.[9]

For clinicians working across cultural lines, there is also a growing literature on cultural competency with several hundred books now available on the subject. For example, Bonder and Martin present a novel methodology for health care workers engaging culturally diverse population,[10] whereas an edited collection on multicultural therapies by Derald Wing Sue and colleagues examines a broad selection of cases, though primarily within the US.[11] Thea James recently offered a series of steps to ensure better quality interactions between emergency room physicians and patients from different cultures. Her general observations can readily be extrapolated to the psychotherapy encounter:

> The definition of culture applies to patients and also physicians. The different cultures present in these two groups affect both parties when they begin a clinical encounter and become the lens through which each party views and interprets the other. Penetrating the barrier created by this 'elephant in the examination room' requires a keen awareness of all the active

elements actually present in the examination room during the patient-physician clinical encounter. These elements include bi-directional culture, social constructs, stereotypes, biases, competence, power and a level of patience. The most important element for physicians, however, is bi-directional culture, which means to understand that both patients AND physicians have a culture that always accompanies them to any joint clinical encounter.[12]

Although this chapter will not primarily attend directly to questions of cultural competency, it should be recognized as a topic in the background. Instead, I will focus on bi-directionality and in particular on the impact cross-cultural analytic encounters have on the clinicians who offer services, and ultimately to the influence this impact may or may not have on "mainstream" traditional Jungian theory and practice—it is too soon yet to know for certain, especially regarding the extent of these influences, but evidence is beginning to emerge as to what/where we might look in the future. As a complement to this approach, I will briefly discuss new voices emerging from DGs and related activities and how their contributions offer new perspectives and engagements.

Eastern Cultural Influences on Jung

The notion of cultural influence is deeply embedded in the life and writing of C. G. Jung as exemplified by his commentaries on numerous texts from cultures distant from that of his origins, as well as his reflections on his various travels to non-Western European locales. While this subject is vast, beyond review here, I would like to briefly note how Jung's engagement with Eastern thought, especially spiritual traditions from India and China, modified and expanded his conceptualization of analytical psychology.

A collection of Jung's writings on the East can be found in the second half of volume 11 of his *Collected Works* (*Psychology and Religion: West and East*), though other articles are scattered throughout the *Collected Works*, e.g., also see volumes 4, 5, 6, 8, 9i & ii, 10, 13, 15, and 18; *Memories, Dreams, Reflections* (*MDR*), and the two volumes of his letters. As early as his *Psychology of the Unconscious* (the original *Wandlungen und Symbole der Libido* of 1912), Jung was already referring to the *Rigveda* and the *Upanishads* to give amplificatory weight to his psychoanalytic interpretations of myths. Then during the tumultuous time of his "confrontation with the unconscious" (1913–1914) he reports having employed yogic practices to calm himself when activated by the psychic material arising, so that he could continue his explorations of the unconscious.[13] Jung's exit from this "Red Book period" was initiated by his synchronistic receipt in 1928 of a Chinese Taoist alchemical manuscript translated into German by his friend the sinologist, Richard Wilhelm—in English, *The Secret of the Golden Flower*.[14] This experience was highly influential in Jung's subsequent turn to alchemy more generally as a psychological system for personality transformation.

By 1930 in his memorial lecture for Wilhelm, Jung discussed the way in which the Western psyche was unconsciously charged with Eastern symbolism, partially in response to Western colonial aggression in the East.[15] In 1932 he gave a series of lectures on Kundalini Yoga in which he analyzed the chakras as the subtle body system of yoga in psychological terms. This system provided him with an enlarged vision of the psyche, having a higher or supra-personal aspect outside consciousness, even taking this as its cosmological starting point.[16] After his travels to India, however, Jung could not accept the metaphysical quality of the Indian philosophical position about a super consciousness beyond the realm of the unconscious. This became one of the pivotal disagreements with a number of scholars from India and led to his withdrawal from further discourse with them.[17] Nevertheless, Jung was also one of the first Western psychotherapists to recognize that certain, apparently strange, inexplicable symptoms occurring in some of his patients could be more accurately examined when reframed as spontaneous awakenings of Kundalini-like energies in their psyches.[18] While Jung was knowledgeable about his subject at the time of these lectures, his familiarity with the material came from reading and discussion with scholars and colleagues. He had not been to India yet; this did not occur until late 1937 (continuing through the winter of 1938) and, as indicated, this would change his relationship to their viewpoint.

Intrigued by yoga philosophy and Tantrism, Jung was also ambivalent, even anxious about Europeans who "carry Eastern ideas and methods over unexamined into our occidental mentality."[19] In his memorial lecture for Wilhelm, he concluded by speaking of how

Wilhelm had "let the European in him slip into the background" and that this sacrifice ultimately cost him his life.[20] Jung identified with the danger he perceived in Wilhelm:

> Wilhelm's problem might also be regarded as a conflict between consciousness and the unconscious, which in his case took the form of a clash between West and East. I believed I understood his situation, since I myself had the same problem as he and knew what it meant to be involved in this conflict.[21]

Without specifying the reference here, Jung seems to be implicitly referring to his own "confrontation with the unconscious" from late 1913. In treating cultural encounters as equivalent to traumatic activations of the unconscious, there is a collapse of boundaries, as between inner and outer world, or self and other. In these circumstances intrapsychic otherness can too readily be projected, bringing temporary relief to the individual but at great cost to all. The unresolved elements from Jung's "confrontation" period seem linked with his general wariness when encountering non-Western cultures generally (whether in Africa; in the US, e.g., with African Americans in St. Elizabeth Hospital and Native Americans at Taos Pueblo; or in India). Perhaps even more basically, it was resonant with his anxiety couched as the danger of psychic infection or *participation mystique* when in collectives or working with regressed clients.

The fear of losing cultural identity, as through unreflective imitation of a foreign other, formed one of the limiting edges in Jung's own psychology. His writings on his travel experiences in *MDR* as well as comments about ethnic and "racial" differences between groups such as found in the essay "Mind and Earth" reveal his cultural biases in these matters.[22] These and related concerns have been raised previously by various authors; for example, Andrew Samuels has offered essential critiques of Jung's political missteps,[23] and J. J. Clarke, while deeply valuing Jung's engagement with Eastern philosophies, does explore his shortcomings and prejudices in this arena.[24]

Thus, while Jung himself did travel and draw upon materials from diverse cultures for his conceptions, specificity of cultural differences, though acknowledged, was not carefully elaborated. Instead the tendency was for universalization, incorporating cultural variations into examples of archetypal patterns. The articulation of a Jungian theory of culture was therefore left to others, a task begun by Joseph Henderson. In the language of Henderson's followers, Singer and Kimbles, Jung seems to be operating out of a "cultural complex," an impediment to genuine engagement.[25] We are left to wonder which aspects of these concerns are implicitly embraced by contemporary Jungians.

Bi-Directionality in Analytic Work across Cultures

Maturation beyond cultural complexes cannot be accomplished solely through psychological analysis, though cultivation of enhanced cultural and sociological sensitivity can be a starting point. After all, it takes years of graduate studies and fieldwork to develop the competencies of a cultural anthropologist. Although this would not be a reasonable requirement for analysts involved in cross-culture training, becoming alert to the dilemmas inherent in such encounters is crucial—hence the growing need for a careful critique of cultural biases in Jung's theories and practices—however, a full exploration of this is well beyond the scope of this short chapter.

More immediately, initiating scrutiny through a bi-directional reflective process with peoples and cultures engaged in study and training in Jungian psychology beyond the traditional institutes is timely. This is already partially underway informally in several areas of the world where local societies have emerged and are taking on the training of individuals in their own or similar cultures, adapting analytical psychology to be relevant within their own context. For example, in Latin America the local societies are finding new and creative ways to work together collaboratively to bring training to those throughout the region who seek it but are too geographically distant to attend one of the local training programs. There are also training groups in geo-cultural regions outside the traditional ones, such as Japan, Korea, South Africa, and even Russia, with many other national and regional groups beginning to form.

For perspective, the early history of the spread of analytical psychology involved individuals traveling to Zürich for analysis with Jung or one of his close associates. Often these individuals would eventually return to their home countries, gradually assembling and building training communities.[26] Thus the dissemination of Jung's psychology consistently began within what was a foreign milieu for the future training analysts, though often with strong allegiance to Jung personally as well as to Zürich. Although there was some bi-directionality as can be gleaned from the correspondence of these analysts with Jung, in truth the weight of influence in the early days was markedly one-sided with Jung often in a supervisory role, answering questions; however, once local societies began to train individuals from within their own cultures on home ground, the importance of their own context became more evident and differing strands of theory and practice evolved,[27] even though the tendency to seek the archetypal at times washed out some cultural reflections.

As the responsibility for training shifted from Jung and his immediate circle to institutes, starting in the late 1940s, accelerating with the birth of the IAAP in 1955, the sharp asymmetry in relationships between trainers and candidates slowly decreased. These shifts were not due solely to changes in analytic culture but reflect much broader social forces in the post-colonial world beginning in the 1950s which gained intensity in subsequent decades. One indicator of this was the evolving vision of candidates-in-training being increasingly seen and treated as professional colleagues, as if of junior status rather than more infantile metaphors.

As we move more deeply into the twenty-first century with increasing awareness, and appreciation of the potential value of diversity in leading to richer, more complex social bodies, it may be useful to reconceive our organizational life in more ecological terms. This approach privileges attention to the network of interconnections among individuals and groups, attending to reciprocal impacts, all of which contribute to the vibrancy or debility of the whole. Asymmetries still abound in ecological models, but the field of interactions, filled with bi-directional influences often hidden or not in plain view, offer insights not available without this perspective. In a striking environmental example, loss of an apex predator, the grey wolf, eliminated from Yellowstone National Park in the 1920s, resulted in significant top-down generated ecological impoverishment. Elk populations unmolested by wolves expanded and began to graze more on Aspen seedlings diminishing their numbers with a rippling cascade of changes unfolding from this, including decreased beaver populations with their dams, resulting in less wetland, in turn altering bird populations, and so on. Reintroduction of the wolf has helped to restore balance, though this is not a straightforward story and the exact details are debated, but the complex network is real.[28] A secondary effect of including an ecological perspective can be to challenge, as in this case, our traditional symbolic understanding of the wolf as an incarnation of evil, chaotic destruction, primitive aggression. Aggressive instincts in the above example are crucial to a healthy, well-functioning environment. By implication ecological values can alter and expand the scope of the transcendent function.

How might this carry over to cross-cultural analytic work? One personal example came when interviewing a group seeking DG status. As standard procedure there is a clinical presentation by a member of the group with discussion among the members facilitated by several IAAP interviewers. During the presentation a dream was reported which included elements not familiar to the interviewers. While the group politely sought interpretive input from the interviewers, the discussion was restrained. Instead of supplying an answer, the interviewers queried the group as to whether the unknown element (for the interviewers) was familiar to the group; did they know a folktale or story with figures in similar situations acting like the one in the dream? After several minutes' hesitation and hushed discussion the room suddenly erupted into an animated offering of cultural narratives where such parallels occurred. The spontaneous amplification from the group brought them solidly into the clinical presentation in a way that had hitherto been unattainable. Not only did the clinical discussion gain greater depth but the interviewers also were affected by the importance of the implicit cultural knowledge that needed to be accessed for real engagement to occur beyond the level of persona.

Several important points regarding cross-cultural work can be gleaned from this example: first, the need to tolerate one's own lack of knowledge in an encounter with a "foreign" other, even when one is present to assess, evaluate, or offer therapeutic assistance. To accomplish this requires an awareness of the limits of one's knowledge and authority, along with the humility to communicate from this position of finitude. Trust in the other's capacity and desire to engage is a corollary, even though this also requires maintaining scrutiny and careful assessment on both sides while attending to the cultural aspects of the transference/counter-transference field. At the edges of this not knowing lies uncertainty which usually cannot be resolved solely through psychological reflection. Some interpersonal discussion of the uncertainties being experienced is required, aided by requests for clarification that can require insider knowledge of cultural, social, linguistic, or body language with nuanced, contextual meanings. Traits such as tolerance, openness, and respect have been found to facilitate communication but always contextualized within a specific cultural milieu. General empathy, while valuable, is insufficient to truly bridge between cultures.

Research has shown that general empathy is multi-factorial (having cognitive and affective components, which are processed in differing ways in the brain with different time scales).[29] Empathy across cultures adds further dimensions to this already complex subject. In 2003 Yu-Wei Wang and colleagues combined notions of general and cultural empathy together with multiculturalism to coin a new term "ethnocultural empathy," which can be defined as those aspects of empathy that are directed towards people from ethnocultural groups other than one's own. They then proceeded to develop a tool to operationalize this idea, the SEE (Scale of Ethnocultural Empathy), initially composed of three primary components:

> We posit that ethnocultural empathy, as a learned ability and a personal trait, can be assessed. Thus, the construct of ethnocultural empathy, as operationalized for this study, is composed of intellectual empathy, empathic emotions, and the communication of those two. Perhaps ethnocultural empathy includes more than three components; however, we base our operationalization on the existing literature, which discusses only these three dimensions.

Intellectual empathy is the ability to understand a racially or ethnically different person's thinking and/or feeling. It is also the ability to perceive the world as the other person does—that is, racial or ethnic perspective-taking ability. The empathic emotions component of ethnocultural empathy is attention to the feeling of a person or persons from another ethnocultural group to the degree that one is able to feel the other's emotional condition from the point of view of that person's racial or ethnic culture. In addition, it refers to a person's emotional response to the emotional display of a person or persons from another ethnocultural group. The communicative empathy component is the expression of ethnocultural empathic thoughts (intellectual empathy) and feelings (empathic emotions) toward members of racial and ethnic groups different from one's own. This component can be expressed through words or actions. Further, we conceptualized ethnocultural empathy as a trait that can be developed over time. Such a conceptualization allows us to measure interpersonal differences and take into account the developmental model of ethnic perspective taking.[30]

As they built and tested their model, Wang, et. al., discovered there were four basic categories that emerged: Empathic Feeling and Expression, Empathic Perspective Taking, Acceptance of Cultural Differences, and Empathic Awareness—to which they noted "the intellectual aspect of ethnocultural empathy may encompass one's perspective taking and awareness toward racial and ethnic differences."[31]

A separate international team lead by Chato Rasoal translated and tested the SEE in Sweden. They confirmed the four principal components for their sample (788 participants), though noted some differences in distribution of responses. These differences were discussed in terms of contemporary Swedish culture which is in the process of becoming more ethnically and culturally diverse.[32] Additional studies by this group also suggest significant overlap between the various forms of empathy.[33] Nevertheless, Italian researchers found the SEE effective with students,[34] while in Turkey researchers favored a three-factor-analysis and found gender and ethnic differences.[35] While there seems to be considerable interest in the SEE, some refinement of the factors which comprise ethnocultural empathy is needed and is being pursued.

To date such measures and concepts have not often been employed by analytical psychologists, though Reeves does refer to ethnocultural transference and countertransference in the therapeutic dyad.[36] An opportunity therefore exists to study the role of ethnocultural dynamics in cross-cultural training. This could effectively be undertaken as a research project focused on bi-directional issues, exploring the cultural projections, expectation, limitations, and discoveries occurring during training by the various individuals involved. Anecdotally, I have noted when there are reciprocal attempts to understand and explore cultural differences in the here-and-now of exchanges, whether analytic, supervisory, or didactic, the alliances between partners improve, at times markedly. At key moments[37] these interactions can lead to a "dyadic expansion of consciousness,"[38] especially when each partner is able to contribute insights from their own perspective. I believe cultural expansion of consciousness could also significantly contribute to a widening of analytical psychology more generally. As those analytic travelers affected by these sorts of encounters write up the ways in which their models have been challenged by their experiences in other cultures, we can collectively gather up these reports and reassess our theories and practices. An initial attempt to do this began with a panel on the topic at the 2012 "2nd European Conference on Analytical Psychology" held in St. Petersburg, Russia, where Jan Wiener, Jorge Rasche, and Shiuya Sara Liuh presented, while I moderated—Jan from London, and Jorge from Berlin, each has served in various roles in IAAP's DG-related activities; Shiuya was a member of the Taipei DG in Taiwan and also has experienced training in Zürich. Their perspectives were all testaments to the powerful impacts these cross-cultural encounters can engender and each pointed to ways their own views were changed in the process. To go further, however, a more explicit research program is needed, e.g., starting with a Jungian-oriented research instrument for assessing bi-directional cultural influences.

Of course there will always be the potential for resistances and inhibitions in cross-cultural exchanges, whether due to fear of rejection, being misunderstood, or shame about the meaning of differences for either or both parties, which are often contingent on unresolved idealizations that can easily arise across cultural gaps. The shadowy projections which break down ethnocultural empathy need conscious attention and exploration. From the perspective of a teacher in these settings it seems one of the most pernicious attitudes that can be taken is dogmatism (often unacknowledged), whether about models of the mind, modes of therapeutic action, or even presumptions about valid goals of treatment (e.g., autonomy for the individual). Within a mono-cultural context claims to truth often can go unreflected especially when basic assumptions are shared between partners; across cultures these implicit beliefs frequently need to be raised to consciousness and examined, often through deliberate questioning of oneself and of one's partner.

New Voices

From the earliest days of analytical psychology individuals from various cultures came to Zürich seeking to engage as students and/or analysands; initially the flow was from the center out. As a broader more polycentric network has evolved through the actions of the IAAP, more people have been able to experience this psychology within the context of their own culture. The analysts who have provided access to the psychology have mostly come from countries with long histories of training. They tend to use translators and to feel the otherness more acutely than when working at home with people from other cultures, though this modulates over time, especially for analysts who see routers by telecommunication between visits. Frequently, analysts do try to learn about the history, culture, economics, religion, art, cuisine, and so on, of the places they visit as well as cultivating a modicum of the dominant language. Sometimes the visiting analysts have ancestral roots in the host culture which can increase their empathic resonances and desires to get to know the cultures they are visiting; others may have long-standing interests in these cultures which enhance their affective links to them. In general, the challenge is for the analyst to integrate herself into the host culture rather than the earlier model of an individual going to a foreign training site where the analysts are at home, and the candidate seeks to adapt at least temporarily to living in that environment—both approaches have benefits and shadows. On-site training creates new possibilities for how analytical psychology is understood when experienced locally, even if a person ultimately completes training at an established institute.

There are two relatively new authors in analytical psychology that I would like to mention whose early encounters with analytical psychology were in the context of the DG program. Both of them have looked at Jung's engagement with their cultures and have used this to understand their own psychology but also have used their cultural knowledge to provide new insights into Jung and his psychology. The first is Shiuya Sara Liuh from the St. Petersburg Conference panel already mentioned. In an article "Chinese Modernity and the Way of Return" she revisits the source of Jung's exit from *The Red Book* and turn towards alchemy through the *Secret of the Golden Flower*.[39] In providing a fuller historical background than is usually given, she is able to show how its transplantation to the West (by Wilhelm and Jung) occurred at a crucial moment in Chinese history. She then offers a post-colonial reading of the text which draws upon Taoist cosmology using the ancient characters for *Hun* (representing "heaven and the yang principle" in the human psyche), *P'o* (representing "earth, body and the yin principle" in the psyche) and *Hui* ("to return or go back to" and "to wind, to circle, to rotate.")[40] In the process she also demonstrates how Jung was able to improve Wilhelm's translation through a non-reductive, psychologically-informed differentiation of gender. In effect, Shiuya Liuh uses Jung's methods combined with her own historical and cultural knowledge to critique his source and to give back a fuller picture of the deep background at play in Jung's use of this Chinese text, enriching our understanding beyond Jung's own statements.

The second author whose work emerged from experiences in a DG is Sulagna Sengupta from Bangalore. In her remarkable *Jung in India*, she recreates Jung's encounter with India, from his philosophical fascination to his extensive trip through India in late 1937 through early 1938; in addition, she discusses his relations with Indians and Indian philosophy.[41] To build a complete picture she has painstakingly combed through numerous archives in India as well as in the West, obtaining access to materials that would have been quite difficult for someone outside the culture to locate (e.g., she speaks a number of Indian languages and dialects). While highly sympathetic to Jung, she does reveal the complexity of his encounter with India and the many unresolved dilemmas and difficulties which arose during and after his trip.

In terms of new directions in analytical psychology coming from the DG program, the innovative projects carried out by Heyong Shen and Gao Lan in China also deserves mention.[42] Their socially-oriented applications of Jungian principles in interventions with populations who have suffered natural disasters and with orphanages have won them national and international acclaim. One of the analysts who worked closely with them in the DG program to develop this approach, Eva Pattis Zoja, has also helped to extend the use of this approach in Latin America.[43] At the IAAP congress in Copenhagen in 2013 Eva, together with Eduardo Carvallo, gave a remarkable presentation entitled "Expressive Sandwork: An Experience Working with Colombian Vulnerable Population."[44] They detailed how they built a team including IAAP analysts and routers to work with displaced, vulnerable populations, offering psychosocial support to children through the medium of sandtray with trained non-interpretive witnesses. Their work eventually involved parents and families in interventions which resulted in greatly enhanced safety for the entire community. There are also numerous other projects underway in various areas of the world but a fuller listing will await a longer publication.

Conclusion

The topic of bi-directionality is timely, perhaps reflective of a shift in focus from exclusive attention on objective knowledge or knowing, to a broader more interactive model of learning. The subjective experience of the one who is learning is now to be included in the image, and the sharper distinctions between teacher and student have softened. Authority is reconfigured, no longer solely a matter of (earned) position but also borne by the way in which information is transmitted; each of us is compelled to consider ourselves as learners and teachers, though in different relative portions. Furthermore, bi-directionality leads into networks and communities, so that ecological formulations help us to grasp more holistic aspects of our encounters.

Training routers, helping to forge new analysts, is a communal project with communal implications and consequences. By wrestling with and embracing the multi-dimensional qualities of otherness in analytic training encounters, a way is opened for fuller psychological engagement. Informational knowledge needs to be tempered with the capacity to tolerate uncertainty and respect for radical otherness. Only such an attitude can unshackle us from dogmatic imitation and release the potential for creative, empathic contact that enhances interconnectedness towards an eco-community of the psyche.

By pursuing greater understanding of bi-directionality, living it as consciously as possible, we generate more profound networks than can be grasped solely from within a dyadic encounter. The complexity that emerges from such activities transcends individual consciousness but leads to greater vibrancy and vitality for those who participate. In a holistic manner the pursuit here is for a social-communal reorganization that grows in richness as the diversity of groups is allowed to expand. In the end the challenge of conscious bi-directionality is whether we can allow ourselves to grow beyond our inherited constraints.

NOTES

1. Georg Northoff, *Philosophy of the Brain: The Brain Problem* (Amsterdam: John Benjamins Publishing, 2004), p. 30.

2. Mark Lyte, "Microbial Endocrinology in the Microbiome-Gut-Brain Axis: How Bacterial Production and Utilization of Neurochemicals Influence Behavior," *PLoS Pathog* 9 (11, 2013): e1003726.

3. Beatrice Beebe and Frank M. Lachmann, *Infant Research and Adult Treatment: Co-Constructing Interactions* (Hillsdale, NJ and London: The Analytic Press, 2002).

4. Linda Carter, "Countertransference and Intersubjectivity," in M. Stein, ed., *Jungian Psychoanalysis* (Chicago, IL: Open Court Press, 2010).

5. Linda Carter, "Amplification as a Form of Cross-Modal Matching in Early Relational Trauma and Disorganized Attachment," in F. Bisagni, ed., *Dialghi D'Infanzia* (Torino, Italy: Anitigone Edizioni, 2010).

6. Linda Carter, "A Jungian Contribution to a Dynamic Systems Understanding of Disorganized Attachment," *Journal of Analytical Psychology* 56 (3, 2011): 334–40.

7. Linda Carter, "Reflections on Bi-Directional Influence in the Matisse/Picasso Relationship and in Clinical Practice from a Complex Adaptive Systems Perspective (CAS)," *Quadrant* 38 (1, 2008).

8. Ed Tronick and Marjorie Beeghly, "Infants' Meaning-Making and the Development of Mental Health Problems," *American Psychologist* 66 (2, 2011): 107–19. doi: 10.1037/a0021631.

9. E. Tronic et al., "Dyadically Expanded States of Consciousness and the Process of Therapeutic Change," *Infant Mental Health Journal* 19 (1998): 290–99.

10. Bette Bonder and Linda Martin, *Culture in Clinical Care: Strategies for Competence,* 2nd ed. (Thorofare, NJ: Slack Inc., 2013).

11. Derald Wing Sue, Miguel E. Gallardo, and Helen A. Neville, eds., *Case Studies in Multicultural Counseling and Therapy* (Hoboken, NJ: John Wiley & Sons, Inc., 2014).

12. Thea James, "The Patient-Physician Clinical Encounter." Accessed 2013, at http://www.med-ed.virginia.edu/courses/culture/pdf/marcuschapter006patientphysicianclinicalencounterrevisedgc.pdf.

13. Sonu Shamdasani, ed., "Introduction" in *The Psychology of Kundalini Yoga: Notes of the Seminar Given in 1932 by C. G. Jung* (Princeton, NJ: Princeton University Press, 1996), p. xxv and n. 30.

14. Joseph Cambray, "*The Red Book*: Entrances and Exits," in *The Red Book: Reflections on C. G. Jung's Liber Novus*, eds. T. Kirsch and G. Hogenson (New York and London: Routledge, 2013).

15. Shamdasani, "Introduction," pp. xvii–xviii and n. 3.

16. Sonu Shamdasani, ed., *The Psychology of Kundalini Yoga: Notes of the Seminar Given in 1932 by C. G. Jung* (Princeton, NJ: Princeton University Press, 1996), pp. 65–67.

17. Sulagna Sengupta, "Jung's Links with India after 1937," in Sulagna Sengupta, *Jung in India* (New Orleans, LA: Spring Journal Books, 2013).

18. Shamdasani, "Introduction," p. xvi and n. 36.

19. *Ibid.*, pp. xxi–xviii and n. 15.

20. C. G. Jung, "Richard Wilhelm: In Memoriam," in *The Collected Works of C. G. Jung*, vol. 15, trans. R. F. C. Hull (London: Routledge and Kegan Paul, 1971), §§ 53–62.

21. Richard Wilhelm, "Appendix IV," in C. G. Jung, *Memories, Dreams, Reflections* (New York: Pantheon Books, 1963), p. 377.

22. C. G. Jung, "Mind and Earth," in *The Collected Works of C. G. Jung*, vol. 10 (Princeton, NJ: Princeton University Press, 1970), §§ 93–103.

23. Andrew Samuels, *The Political Psyche* (London and New York: Routledge, 1993).

24. J. J. Clarke, *Jung and Eastern Thought* (London and New York: Routledge, 1994); see especially chap. 9, pp. 158–78.

25. Thomas Singer and Samuel Kimbles, eds., *The Cultural Complex* (Hove and New York: Bruner-Routledge, 2004); also see their chapter "The Emerging Theory of Cultural Complexes," in *Analytical Psychology:*

Contemporary Perspectives in Jungian Analysis, eds. J. Cambray and L. Carter (Hove and New York: Bruner-Routledge, 2004), pp. 176–203.

26. See Thomas Kirsch, *The Jungians* (London and Philadelphia: Routledge, 2000) for a more in depth history of the early development of analytical psychology.

27. By the early 1980s at least three separate schools of analytical psychology had formed. The development of these different approaches together with a comparative analysis was well portrayed in Andrew Samuels' book *Jung and the Post-Jungians* (London and Boston: Routledge and Kegan Paul, 1985).

28. Emily Gertz, "Has the Reintroduction of Wolves Really Saved Yellowstone?" *Popular Science*, Mar. 14, 2014. Accessed Aug. 9, 2014, at http://www.nps.gov/yell/naturescience/wolfrest.htm; and see "Wolf Restoration Continued," National Park Service. Accessed Aug. 9, 2014, at http://www.popsci.com/article/science/have-wolves-really-saved-yellowstone.

29. For a brief review of this see chap. 4 of Joseph Cambray, *Synchronicity* (College Station, TX: Texas A&M University Press, 2009).

30. Yu-Wei Wang et al., "The Scale of Ethnocultural Empathy: Development, Validation, and Reliability," *Journal of Counseling Psychology* 50 (2, 2003): 222.

31. *Ibid.*, p. 230.

32. Chato Rasoal et al., "Development of a Swedish Version of the Scale of Ethnocultural Empathy," *Psychology* 2 (6, 2011): 568–73. doi: 10.4236/psych. 2011. 26087.

33. Chato Rasoal et al., "Ethnocultural versus Basic Empathy: Same or Different?," *Psychology* 2 (9, 2011): 925–30.

34. Paolo Albiero and Giada Matricardi, "Empathy Towards People of Different Race and Ethnicity: Further Empirical Evidence for the Scale of Ethnocultural Empathy," *International Journal of Intercultural Relations* 37 (5, 2013): 648–55.

35. Gözde Özdikmenli-Demir and Serdar Demir, "Testing the Psychometric Properties of the Scale of Ethnocultural Empathy in Turkey," *Measurement and Evaluation in Counseling and Development* 47 (1, 2014): 27–42.

36. Kenneth M. Reeves, "Racism and Projection of the Shadow," *Psychotherapy: Theory, Research, Practice, Training* 37 (1, 2000): 80–88.

37. For a discussion of transformation moments see Joseph Cambray, "Moments of Complexity and Enigmatic Action: A Jungian View of the Therapeutic Field," *Journal of Analytical Psychology* 56 (2, 2011): 296–309.

38. Tronick et al., "Dyadically Expanded States of Consciousness and the Process of Therapeutic Change."

39. Shiuya Sara Liuh, "Chinese Modernity and the Way of Return," in *How and Why We Still Read Jung*, eds. Jean Kirsch and Murray Stein (London and New York: Routledge, 2013), pp. 144–60.

40. Liuh, "Chinese Modernity and the Way of Return," pp. 152, 156–57.

41. Sengupta, *Jung in India*.

42. See, e.g., Heyong Shen and Gao Lan, "The Garden of the Heart and Soul: Psychological Relief Work in the Earthquake Zones and Orphanages in China," in *Spring 88: Environmental Disasters and Collective Trauma* (New Orleans, LA: Spring Journal Books, 2012), pp. 61–73.

43. Eva Pattis Zoja, "Expressive Sandwork: A Jungian Approach in Disaster Areas," in *Spring 88: Environmental Disasters and Collective Trauma* (New Orleans, LA: Spring Journal Books, 2012), pp. 75–90.

44. Eduardo Carvallo and Eva Pattis Zoja, "Expressive Sandwork: An Experience Working with [a] Colombian Vulnerable Population," in Emilija Kiehl, ed., *Copenhagen 2013—100 Years on: Origins, Innovations and Controversies (Einsiedeln: Daimon Verlag, 2014)*.

CHAPTER 3

THE SELF AND INDIVIDUATION
UNIVERSAL AND PARTICULAR

ANN SHEARER

The Self and individuation: what two concepts could be more central to the Jungian tradition? Anyone who picks up the standard handbook will discover that "individuation is perhaps Jung's major psychological idea, a sort of backbone for the rest of the corpus."[1] And there is no individuation without the Self as both guide and goal, the ego-Self axis as orientation. Individuation's work of bringing together disparate psychic fragments is only in part an act of ego's will, says Jung: "in another sense it is a spontaneous manifestation of the self, which was always there."[2] Yet perhaps just because these concepts are so central, they continue to generate questioning and debate. By exploring them in cultural contexts very different to that of their conception, this chapter hopes to bring another perspective again to questions about their nature and energy.

Jung himself left generous room for debate. Individuation is not a therapy, he said, but simply that "biological process ... by which every living being becomes what it was destined to become from the beginning."[3] But then again, it is "a relatively rare occurrence, which is experienced only by those who have gone through the wearisome ... business of coming to terms with the unconscious components of the personality."[4] By the end of his life, in the protocols that became bowdlerised into his "autobiography," Jung saw it as "a lonely search, perhaps akin to the process of dying."[5] And for all his assertions of the Self as the principle and archetype of orientation and meaning, both centre and circumference of psyche's wholeness, he also knew it to be no more than a construct to express the unknowable, a "dream of totality."[6]

So the questions multiply—indication enough of the lasting pull of the idea. Does the Self even exist, or is it just an "old monotheistic structure of unity and centering?"[7] Is it an idea of psychic totality or an archetype within that totality, goal, process, or both, pre-existent or emergent? Or even "[t]he totality of our action in the world ... of everything from the cell division of the embryo ... to writing a symphony or committing mass murder?"[8] And, most persistently of all, is it written with an upper or lower case *S*? This last question has vexed commentators for at least thirty years now.[9] "One does become weary dealing with the question of whether the term self should be capitalised or not," sighs Murray Stein; he himself opts for the lower case, lest there be confusion with Divinity.[10] The scope of the Self (my own preference, lest there be confusion with ego-identity) remains, it seems, immeasurable.

The idea of individuation has perhaps attracted less debate. Jungians generally have seemed more willing to accept it at Jung's own estimation, as the goal of human endeavour. But what does it really mean? Vague references to "fulfilling one's potential" and "becoming the person one is meant to be" can sound a great deal cosier than Jung's sense of weary loneliness, his insistence that the all-encompassing Self must inevitably be both bright and dark.[11] They take little account either of the complexity of relationship between individuation and the specific collective in which individuals are embedded, or of the intellectual, social, and economic realities through which they are also formed. Jungian tradition makes much of Jung's dislike of the group and mistrust of "the mass"; if individuation has come to mean anything, it is the liberation of the individual from unconscious conformity with social norms and expectations. But inevitably this opposition is coloured by five hundred years and more of cultural and intellectual history: the "development" of the individual has since the Renaissance been a defining and often glorious story of Western European culture and others that have grown from its values. For even its relatively near neighbours, the story may be very different. I remember the freezing *frisson* that went through the room on my first teaching visit to St. Petersburg when I heedlessly mentioned Jung's ideas about the psychological relationship between individual and collective: in countries of the former Soviet Union, the juxtaposition of those two words must surely carry a historical burden of personal and cultural fear, anger, and grief which is barely imaginable to outsiders.

For others again, a secure embeddedness in the collective may be an inalienable part of individual identity, which is denied at great personal and cultural cost. Anjali D'Souza explores this in her essay on the personal and collective trauma of the brutal Portuguese Christianisation of Goa, still experienced even 500 years later. The converts were not just taken from their religion, for "Hinduism" so identified was a colonial construct. They were torn from their deepest Hindu identity, which is rooted in a living relationship to the land and its ancient histories and myths and carried through the minutest details of social mores and behaviours, of music, food, art, and physical expression. This deracination still painfully reverberates in individual and cultural dissociations. But the deep identity remains in the unconscious, and its return to consciousness may bring healing reconnections for individuals and perhaps for Goa itself.[12]

In her evocation of her adoptive city of Bangalore, Kusum Dhar Prabhu shows just how lastingly alive these connections may be. A new sort of colonialism has torn down and rebuilt the city into a by-word for call-centered "modernity." But the life of the "other Bangalore" is still held in Shiva's cosmic dance and its inhabitants' whispered supplications into his bull Nandi's ear; the threads tied round sacred trees still weave the containment of the Mother Goddess whose bounty is encoded in the derivation of Bengaluru, the city's "real" name. In India, it came as a shock when Bangalore became the capital of the first state to return a Hinducentric BJP government. But Dhar Prabhu sees this as a longing of the "modern" to reconnect to religious roots. Each needs the other: "Indian teachings tell us that, as mortals, we carry within us an immortal, transpersonal nucleus, the Self. ... As Indians, we are always asked to live in two worlds."[13]

Importantly, the logic of Jung's own theory sees no rupture between individual and collective: both are equally fuelled by the same archetypal energies of the collective unconscious. In this understanding, the energies and actions of the Self may manifest in similar ways at cellular, individual, group, and societal levels.[14] The old myths knew this too. In them, there is always something beyond the heroic "individuation journey" of separation from society and its culmination in the union of opposites: the journey is only over when the hero comes home to the collective to establish his kingdom.

Jung knew by 1916 that "both the identification with the collective and the will to segregate oneself from it are alike synonymous with disease."[15] Two years later, he wrote of the guilt which "the individuant" [himself?] can only redeem by returning "values" to their society.[16] His eventual solution to the problematic interplay between individual and collective remained constant. "Only a change in the attitude of the individual can bring about a renewal in the spirit of the nations," he wrote in the "general running amok" that followed World War One.[17] And forty years on he still held to that: "the salvation of the world consists in the salvation of the individual soul."[18]

Well, does it? James Hillman for one latterly moved dramatically away from his long-held concern with individual soul-making, to talk about "the individuation of each moment in life, each action, each relationship, and each thing ... so that the innate dignity, beauty, and integrity of any act and any thing from doorknob to deck chair to bed sheet may become fully present in its uniqueness."[19] For Wolfgang Giegerich, too, the very project of individuation is psychologically obsolete, however rewarding it may be for individuals. But Hillman's devotion to *anima mundi* is for him simply nostalgia. The real *magnum opus* of the soul in this age, the real work of objective psyche, is globalisation; the supreme contemporary value is maximising profit, and individual humans are only of interest for their value to the production process.[20] This bleak view of individual insignificance is far from Andrew Samuels' concern with the "political psyche," his project "to make every citizen into a potential therapist of the world" a "psychological citizen" whose individuation must include political awareness.[21]

Wherever we stand on this spectrum, one thing seems clear. Individuation can only happen in relationship—whether with Hillman's doorknobs and deckchairs, Samuels' disadvantaged groups, or even Giegerich's harsh forces of globalisation. Jung said it first: "[I]ndividuation is an internal and subjective process of integration [a]nd ... an equally indispensable process of objective relationship. Neither can exist without the other."[22] In a posthumous paper, Louis Zinkin takes on the logic to assert that the Self too can come into existence only through interaction with others. "The self is always a construction, one which is not possible without language and language is not possible without culture and culture is always shared."[23]

Through these different understandings, the very concepts of the Self and individuation move from the universal to the particular, from givens of human nature to cultural constructs, from the essentially unchanging to a variety of expressions in different mundane realities. This opposition, or rather complementarity, seems to have been at the heart of my work and learning with Jungian psychology over more than a decade in Russia, and also elsewhere.

About halfway through four years as a visiting supervisor in St. Petersburg, I characterised the work as both "strange and familiar" and so it seems still.[24] This sense of universal similarities of the human psyche and the sometimes extraordinary "strangeness" of their cultural context has been part of my experience and education in South Africa, India, Estonia, and Bulgaria as well.

The tag alone does nothing to capture the intensity of these experiences. It says nothing about the times of delight and even wonder that people so different in their personal experience, history, and culture could so deeply enjoy working together, or the ricochets into bewilderment, anger, and even despair that I would ever understand them—or they me. In Renos Papadopoulos's useful distinction, these "Others" seemed sometimes "familiar," sometimes "exotic."[25] The idealisations of the second state may be exhilarating, but these too can collapse into their negative opposites. And where violence is an ever-present local reality, then this "otherness" is indeed a world away from my own protected, white middle-class habitat.

So these experiences have been both strange and familiar to me in ways far beyond the superficial and inevitably intensified again by the individual and cultural complexes that have been constellated between us. Complexes too are expressions of the Self's psychic totality, and consulting room experience tells us just how much of the work of individuation consists in understanding how they shape us. When the complexes are cultural the individual and the collective are by definition both affected, and I have an abiding sense that while we were ostensibly working for our personal selves, something larger was also at play.

Both "we" and "our" seem important here. As formulated, the "cultural complex" is based on repetitive and historical experiences that have taken root in the cultural unconscious of a group and manifest in the life of individuals and the group itself. "Mostly these complexes have to do with trauma, discrimination, feelings of oppression, and inferiority at the hands of another offending group—*although the "offending groups" are just as frequently feeling discriminated against and treated unfairly.*[26] (my italics) So the cultural complex is something shared. This mutuality seems to have been relatively little explored. Writings on cultural complexes have emphasised their effects on the traumatised rather than the traumatisers—perhaps inevitably so in individual case studies, but in historical accounts of collective sufferings as well.[27] Yet in retrospect at least, it is the shared nature of the complexes that illuminates much of my experience of working in different cultures—and of the ways of individuation and the Self through these.

When I was preparing my first teaching weekend in St. Petersburg, back in 2000, three memories came unbidden and insistently to mind. The first was of my seventeen-year-old self in Geneva and a reception to which a family friend had invited a visiting Russian delegation. Russians! A party! This juxtaposition felt profoundly wrong to me, and I helped set out the little cakes and vodka in a confusion of excitement and dread. Nothing in my Cold War cultural upbringing had given a hint of the charm with which the visitors accepted my fearful waitressing, melting my apprehension into fleeting wordless flirtations; nothing had allowed for "the Soviets" to be other than the enemy followers of a morally repugnant creed. The other two memories came from my only visit to the Soviet Union, some ten years later. Straightway on arrival in Moscow, I got it wrong as I offered our unsmiling guide—oh how naively!—my newspaper for her possible interest. Nothing but lies and propaganda, she barked, like all Western journalism. And then, there was our young local guide at the war memorial in Kiev and her movingly heartfelt little speech about the importance of our peoples coming together in understanding so that one day the horrors of war would finally cease. She never returned; the next day our still-unsmiling group guide took over.

So I inevitably brought questions to that first St. Petersburg teaching weekend—about how we would be able to communicate, about how my "guidance" would be received, about whether I too would be dismissed. But more than that, I took a certain ingrained sense of the moral superiority of "my" open, democratic history over "their" repressive one—and by extension of my Jungian understandings over their psychological unawareness. In the many visits that unfolded from that first one I could certainly too often sense in myself that unsmiling group guide waiting defensively to pounce if I felt that my version of the Truth was not respected. But even she was not protection enough against feelings of inadequacy as I struggled to understand so different a cultural and individual experience; there were times when these people seemed *older* than I was, though they were in fact all younger. Vladimir Tsivinsky has written of the Russian cultural complex as spanning the poles between grandiosity and inferiority and the deep split between these poles as leaving his patients in a "no-place" of longing for the Utopia that the Soviet Union once promised and still for many represents.[28] That swing between superiority and inferiority was encoded in my experience too, and I now recognise it as operating between our groups as well as within us as individuals, as fuel for a cultural complex which we shared.

The swing has been brought painfully back home to me more recently. Towards the end of 2013 it was reported that the UK and other EU countries would lift the unprecedented controls that had restricted the working rights of Bulgarian and Romanian citizens since their two countries had joined the Union. The decision unleashed a media-fuelled panic that Britain's infrastructure would be swamped by a tide of migrant scroungers; an e-petition to the Home Office to renew controls drew over 150,000 signatures. This xenophobic caricature could not be further from my pleasure in working in Bulgaria over four years now with people so deeply engaged with Jungian psychology, despite an extremely difficult political and economic reality (or further, as it happened, from the meticulous attentions of a team of Romanian builders in my house). Each time the negative public reaction came up I felt a wave of anger and shame at the presumed "superiority" of the British over these "inferior" nations. I thought back to my first visit to Sofia and learning at the National Museum something of Bulgaria's pride in its history. Then suddenly I had a flash memory of schoolgirl boredom at Gladstone's campaign to publicise the appalling atrocities with which the Ottomans crushed the 1876 Bulgarian rebellion. At least we got a snigger when Gladstone declared "The Bulgars are openly revolting."

Superiority, inferiority: the energy of these archetypal poles is perhaps inevitably constellated in any teacher-student relationship, and the cultural complex can only add intensity. Catherine Crowther and Jan Wiener, for instance, have written of their difficulties as organisers of the International Association for Analytical Psychology (IAAP) training in Russia, and of how, particularly because they were women, they found their authority as leaders both idealised and denigrated.[29] As a supervisor, though mercifully free of organisational responsibility, I too was authorised by the IAAP to bring the authority of my professional experience to a relationship that has its own intensities. As John Beebe has pointed out, few people come to analysis in search of a new identity; mostly, they want to ameliorate the one they have. Yet a new identity as analyst is precisely the goal of students who come to supervision, and this gives the supervisory relationship a particular emotional charge.[30] Supervisees are seeking their own *authority* in a sense that goes back to its Latin roots: they are working, however consciously or unconsciously, to become *authors* of their own psychotherapeutic selves—surely a work of individuation. This is what the supervisor is there to foster. But sharing authority is never easy. Crowther and Wiener write of the loss of safety, narcissistic gratifications, and inflations that went with their own movement from hierarchical to distributed leadership.[31] "Authority knows how to mourn the death of its own projects for the other," says the theologian Jean Vanier.[32] Not surprising, perhaps, that authority's power shadow is waiting to erupt.

The nature of training invites this. Students are seeking not only an inner authority but more straightforwardly the one conferred by diplomas and society membership; teachers are also judges and gatekeepers. Particularly, perhaps, where issues of authority have been so personally and collectively painful, power can seem very present. We well-meaning visitors maybe underestimate how intimidating we can appear. Tatiana Rebeko, for instance, recalls that it was only eighteen months into her IAAP Router Training Programme that she and fellow students stopped being afraid of their teachers and ashamed to ask about what they didn't understand; only then could they feel gratitude for the training as gift rather than experiencing it as charity.[33] Alena Tserashchuk looked back on feelings of being studied and scrutinised as if under a microscope and perceived as belonging to a "primitive," even animal, population. Vladimir Tsivinsky anticipated that his interview for the programme would be like a police interrogation, one of the "trials" he had to go through, like meeting the "strict and arrogant" British lady analyst of his fantasies. If in retrospect they could see these perceptions as individual "paranoid–schizoid defences" and projections of "internal idealized persecuting objects," [34] for me, that detracts nothing from the experienced power of the IAAP and its agents. Nor, I think, does it diminish the reality of the power-complex. I could certainly feel this tingeing my own work. For all the care that was taken to draw the UK training team for Russia from each of the then four IAAP organisations in London, one school of thought—the developmental—was dominant. I had to be mindful that my efforts to hold the authority of my own classical Jungian approach didn't shade into too powerful a competitive desire to have the students see it my way rather than encouraging them to find their own.

Looking back now, these issues of power and authority, which are so much part of the processes of individuation, seem to have been played out above all in struggles around boundary and frame. In 2002, Tatiana Rudakova was already remembering the often unvoiced but still discernible "contemptuous surprise" of visiting Western professionals to the East European Institute of Psychoanalysis in St. Petersburg in the early 1990s: "How dare you break the boundaries and call what you do psychoanalysis?"[35] Some twenty years later, Catherine Crowther and Jan Wiener recalled how startling they'd found the flexibility of Russian practitioners around the issues of money, time, confidentiality, and holidays, and how uneasy they were about what seemed to them "casual, easy, alterations to the frame." They had to learn about some of the economic realities behind these and to re-examine their own theoretical and clinical assumptions as well. (But at the end of the day, they were able to say, "If there is anything we have left behind in Russia it is a respect for boundaries!")[36]

As a supervisor, I too had my moments of amazement at how far people would go to accommodate their patients, and I too struggled to impart our boundaried Western ways. But the real battleground was the tea room. The timetable for our work was very tight, carefully designed, and signalled in advance to ensure the most time possible for each supervisee. And at the start of the day, they drifted in, made a cup of tea, and chatted to friends until they were good and ready to begin. It drove me to distraction. I tried to see it as an expression of the general boundary issues—and indeed as people learned more about these clinically I finally learned to laugh about "Russian time" and "British time" and we found a tea break compromise. But there was also a power story here, and in retrospect I see in this my own failure to acknowledge the authority which the Russians were struggling to establish.

For people who grew up in the borderless "Soviet nation,"[37] perhaps living in communal flats, learning to observe clinical boundaries may indeed have not just professional but personal importance. For many of their patients too, finding a secure frame and boundaries may be central. Elena Pourtova, for instance, notes how time and again her patients are struggling to establish their "own place"; whatever the individual housing problem, there is always also a question of individuation.[38] But I remain grateful to my Russian supervisees and colleagues for teaching me to question the detailed concern for the theoretical and clinical frame which is such a feature of "post-Jungian" psychology. These days, I see a danger of something defensive in it, a retreat from Jung's understanding of analysis as "real relationship" beyond the transference. Angela Connolly detected this defensiveness in herself when she began to live and work as an analyst in Moscow. Her intense personal experience of disorientation and destructuring led to withdrawal and isolation; in her work she became somewhat obsessively insistent on rules and boundaries, defensively using interpretations to protect her boundaries rather than to foster real insight.[39] There is something collective here too, which I hear in references to "the West" in British and North American Jungian writings, even though the individual cultures which make this up are no more homogenous than were those of the Soviet Union. Perhaps we Jungians have needed a power bloc of our own.

Russian reactions to our analytic ways have often made me think about what makes a therapeutic relationship truly "real" and healing. "Please, do not read!" their students told Catherine Crowther and Jan Wiener when they first started teaching Jungian theory in St. Petersburg; they had to realise that they had been dealing with the unknown by hiding behind the safety of their prepared papers.[40] Some of Angela Connolly's analysands found her "powerful, distant, cold and rigid" and compared her unfavourably with their warm and sympathetic Russian analysts; they told her how difficult it was for her to understand Russian culture, or that analysis wasn't really suitable here. I think my supervisees were telling me through the battle of the tea break that for them professional boundaries should never crush the fluidities of "real" exchange and relationship. Our Russian colleagues were practising psychotherapy before we visitors arrived and they will continue to practice and teach it long after we are gone. I remember Jung's dictum: "Every therapist not only has his own method—he himself is that method."[41]

Experiences of individuation and the Self must be intensely personal and even private, coloured by both individual and cultural experience. Indians live naturally in the two worlds of mortality and the imperishable Self, but very many people who lived under Soviet values suffered damage to the ego-self axis.[42] Over and above the differences, however, I know there have been times in my work when *something happens*. We seem to meet at a new level, one we recognise without discussion. Importantly, this has nothing to do with smoothing over the very real differences between us; I have actually felt more intensely "myself." The experience is akin perhaps to what Catherine Crowther and Martin Schmidt characterise as "states of grace" in analysis—events that happen rather than being deliberately created, when there is a heightened sense of relatedness, a profound transformation of feeling-tone, and a transcendence of impasse.[43] For me, these times of extraordinary, even joyful, intensity are manifestations of the Self as it brings together and contains disparate contents, as much in the group as in and between individuals. Ego's intent is transcended here: these experiences have certainly never happened when we were actually discussing individuation or the Self. (In fact, I remember a weekend seminar on the latter that was more than usually marked by unannounced comings and goings among the participants, a sense of splitting apart rather than coming together.) Most often, the experiences have arrived when we were untrammelled by the ego/persona demands of official training, allowing the symbolic languages of image and strange-yet-familiar mythic tales to carry us beyond the consciously known to recognise a shared substratum.

The Self's bringing together of the disparate and divided may also, as Jung insisted, be deeply painful, and when the mutual cultural complex is touched, the pain can be intensified. I learned something about this unforgettably in Bangalore, when at the end of a lively series of seminars we turned to discussing that concept, initially through the "neutral" example of the lasting effects on the Mexican psyche of the Spanish conquest of the Aztecs.[44] Nothing had prepared me for the suffering and anger that was then unleashed. Stories of shockingly violent dislocation at the time of Partition still carried an enduring grief of individual and collective loss. The Raj still bruised the lives of those unborn until decades after it ended. One young woman told of how as a little girl she'd peered through the hedge at the newly-arrived British family next door, longing to meet these curious and different children. One day, their mother noticed her, and instead of inviting her in, screamed at her to get away and never come near them again. The young woman recounted how her bewildered shock turned to deep shame (What had she done wrong to make the woman so angry with her?) and finally to anger (It's MY country, not hers!).

As these stories and others unfolded—and for some there were no words, just tears—I felt more and more crushed. All our earlier coming together was thrown into question; I found it painful to wonder whether our apparent enjoyment had perhaps been no more than a polite persona for a foreign guest. When at the end of the session someone asked how I felt, I could say only "I feel silenced." The experience left me shocked and distressed about myself and my relationship to India, which has been my loved "exotic Other" for nearly fifty years. I felt I'd been ridiculously naive to think this contemporary coming together could be otherwise than complex and painful, weighted down by layers of history. It was only the next morning that it came to me how uncannily my feelings mirrored those of that little girl. I had come with an open curiosity, eager to make contact with these "Others." Their reaction had thrown me into deep shame, an inherited colonial guilt. And from there, a real anger: "This is not my story—I wasn't part of that—I was a tiny child in 1947—I have a right to be seen for myself!" And like that little girl, I too had felt silenced and unheard.

Yet I remember that experience not with pain, but as one of privilege and some personal healing. Partly, that was because I was finally able to speak of my feelings and to have these acknowledged. But most importantly, I think it was because we were actually drawn closer by speaking of the complex and learning something of its shared nature. The experience has stayed with me through two subsequent, and I hope deeper, visits to Bangalore and I feel I will continue to learn from it—

not just about "Britain and India" but about the processes of the Self and individuation as cultures come together.

In Bangalore's "gentle mixture of creativity and conflict," its citizens have a motto and mantra: "*Solpa adjust madi,*" they say to each other, "Please adjust a little."[45] Not a bad motto either, perhaps, in the mixture of creativity and conflict that characterises cross-cultural studies of Jungian psychology. Right back at the start of the IAAP programme in St. Petersburg, Tatiana Rudakova wrote of the importance of accepting diversity in a professional setting which had already drawn from an influx of visiting teachers of different creeds.[46] As Jungian visitors added their own diversities, Vladimir Tsivinsky for one found this hard to accept. But, he says, he gradually drew on the spectrum to form his own professional approach.[47] He found his own authority.

As one of those visitors, in that setting and in the others too, I know that people with very disparate personal and cultural histories and present circumstances can indeed come together around Jung's ideas, and in the process discover more about living with their own and other people's—or even peoples'—diversities. Jung's old "dream of totality" still seems to be at work, both within and between individuals. Who knows how far the ripples may reach?

Notes

1. Murray Stein, "Individuation," in R. Papadopoulos, ed., *The Handbook of Jungian Psychology* (Hove: Routledge, 2006), p. 196.

2. C. G. Jung, "Transformation Symbolism in the Mass" (1954), in *The Collected Works of C. G. Jung*, vol. 11, ed. and trans. Gerhard Adler and R. F. C. Hull, 2nd ed. (London: Routledge and Kegan Paul, 1969), § 400.

3. Jung, Forward to Victor White, "God and the Unconscious" (1952), CW 11, § 460.

4. C. G. Jung, "On the Nature of the Psyche" (1954), in *The Collected Works of C. G. Jung*, vol. 8, ed. and trans. Gerhard Adler and R. F. C. Hull, 2nd ed. (London: Routledge and Kegan Paul, 1969), § 430.

5. Deirdre Bair, *Jung: A Biography* (Boston, MA: Little, Brown, 2003), p. 429.

6. Miguel Serrano, *C. G. Jung and Hermann Hesse: A Record of Two Friendships* (New York: Schocken, 1966), p. 50.

7. James Hillman, *Inter Views* (New York: Harper and Row, 1983), p. 83.

8. Warren Colman, "On Being, Knowing and Having a Self," *Journal of Analytical Psychology* 53 (3, 2008): 352.

9. Judith Hubback, "Concepts of the Self: The self or the Self?" *Journal of Analytical Psychology* 30 (1985): 229–71.

10. Murray Stein, "Divinity Expresses the Self: An Investigation," *Journal of Analytical Psychology* 53 (3, 2008): 305.

11. C. G. Jung, "Flying Saucers: A Modern Myth" (1959), in *The Collected Works of C. G. Jung*, vol. 10, ed. and trans. Gerhard Adler and R. F. C. Hull, 2nd ed. (London: Routledge and Kegan Paul, 1970), § 640.

12. Anjali D'Souza, "The Ancestral Soul; Fractures and Continuities—Psyche, Memory and the Cultural Complex" (IAAP Intermediate Router paper, 2013).

13. Kusum Dhar Prabhu, "Whispers in a Bull's Ear: The Natural Soul of Bangalore," in T. Singer, ed., *Psyche and the City* (New Orleans, LA: Spring Journal Books, 2010), p. 376.

14. Pamela Donleavy and Ann Shearer, *From Ancient Myth to Modern Healing: Themis, Goddess of Heart-Soul, Justice and Reconciliation* (London: Routledge, 2008); and Ann Shearer, "Myth of Themis and Jung's Concept of the Self," in L. Huskinson, ed., *Dreaming the Myth Onwards* (London: Routledge, 2008), pp. 46–57.

15. C. G. Jung, "The Structure of the Unconscious" (1916), in *The Collected Works of C. G. Jung*, vol. 7, ed. and trans. Gerhard Adler and R. F. C. Hull (London: Routledge and Kegan Paul, 1953), § 485.

16. C. G. Jung, "Adaptation, Individuation, Collectivity" (1918), in *The Collected Works of C. G. Jung*, vol. 18, ed. and trans. Gerhard Adler and R. F. C. Hull (London: Routledge and Kegan Paul, 1977), §§ 1095–98.

17. Jung, "The Role of the Unconscious" (1918), CW 10, § 45.

18. Jung, "The Undiscovered Self" (1957), CW 10, § 536.

19. James Hillman and Michael Ventura, *We've Had a Hundred Years of Psychotherapy and the World's Getting Worse* (San Francisco, CA: HarperOne, 1993), p. 52.

20. Wolfgang Giegerich, "The Opposition of 'Individual' and 'Collective'—Psychology's Basic Fault," Guild Lecture No. 259 (London: Guild of Pastoral Psychology, 1997).

21. Andrew Samuels, *The Political Psyche* (London: Routledge, 1993); "Political, Social and Cultural Material," in R. Withers, ed., *Controversies in Analytical Psychology* (Hove: Brunner-Routledge, 2003), p. 135.

22. C. G. Jung, "The Psychology of the Transference" (1946), in *The Collected Works of C. G. Jung*, vol. 16, ed. and trans. Gerhard Adler and R. F. C. Hull, 2nd ed. (London: Routledge and Kegan Paul, 1966), § 448.

23. Louis Zinkin, "Your Self: Did You Find It or Did You Make It?" *Journal of Analytical Psychology* 53 (3, 2008): 404.

24. Ann Shearer, "The St. Petersburg Experience: Supervision Strange and Familiar," in L. Cowan, ed., *Barcelona 04*: *Proceedings of the Sixteenth International Congress for Analytical Psychology* (Einsiedeln: Daimon, 2006), pp. 634–39.

25. Renos Papadopoulos, "The Other Other: When the Exotic Other Subjugates the Familiar Other," *Journal of Analytical Psychology* 47 (2002): 163–88.

26. Thomas Singer and Samuel L. Kimbles, eds., *The Cultural Complex* (Hove: Brunner-Routledge, 2004), p. 7.

27. *Ibid*.; and Gražina Gudaitė and Murray Stein, eds., *Confronting Cultural Trauma: Jungian Approaches to Understanding and Healing* (New Orleans, LA: Spring Journal Books, 2014).

28. Vladimir Tsivinsky, "The Spatial Metaphor of Utopia in Russian Culture and in Analysis," *Journal of Analytical Psychology*, 59 (2014): 47–59.

29. Catherine Crowther and Jan Wiener, "Images of Authority: Implications for Personal and National Identity" (paper, First European Conference of Analytical Psychology, Vilnius, June 27, 2009).

30. John Beebe, "Some Special Features of the Transference in Consultation and Supervision," in M. A. Mattoon, ed., *Florence 98: Proceedings of the Fourteenth International Congress for Analytical Psychology* (Einsiedeln: Daimon, 1999), pp. 482–89.

31. Crowther and Wiener, "Images of Authority," p. 4.

32. Jean Vanier, *Signs of the Times: Seven Paths of Hope for a Troubled World* (London: Darton, Longman and Todd, 2013), p. 81.

33. Tatiana Rebeko, "Developing Analytic Identity in Different Cultures" (paper, First European Conference of Analytical Psychology, Vilnius, June 27, 2009).

34. Alena Tserashchuk, "Untitled," in Catherine Crowther et al., "Fifteen Minute Stories about Training," *Journal of Analytical Psychology* 56: (2011): 629–34; and Vladimir Tsivinsky, "Impressions and Reflections," in Crowther et al., "Fifteen Minute Stories about Training," pp. 634–38.

35. Tatiana Rudakova, "Diversity of Learning Psychoanalysis and Analytical Psychology," *Journal of Analytical Psychology* 47 (2002): 302.

36. Catherine Crowther and Jan Wiener, "Reflections on Cross-Cultural Training in Russia" (lecture, Bowlby Centre, London, May 12, 2013).

37. Tserashchuk, "Untitled," p. 629.

38. Elena Pourtova, "Moscow is Like a Sweet Berry," in Thomas Singer, ed., *Psyche and the City* (New Orleans, LA: Spring Journal Books, 2010), pp. 195–96.

39. Angela Connelly, "Through the Iron Curtain: Analytical Space in Post-Soviet Russia," *Journal of Analytical Psychology* 51 (3, 2006): 173–91.

40. Catherine Crowther and Jan Wiener, "Finding the Space between East and West: The Emotional Impact of Teaching in St. Petersburg," *Journal of Analytical Psychology* 47 (2, 2002): 285–300.

41. Jung, "Medicine and Psychotherapy" (1945), CW 16, § 198.

42. Vsevolod Kalinenko and Madina Slutskaya, "'Father of the People' versus 'Enemies of the People': A Split-Father Complex as Foundation for Collective Trauma in Russia," in Gudaitė and Stein, *Confronting Cultural Trauma*, pp. 95–112; and Ursula Peterson and Monika Luik, "Expressions of Transgenerational Trauma in the Estonian Context," in Gudaitė and Stein, *Confronting Cultural Trauma*, pp. 193–210.

43. Catherine Crowther and Martin Schmidt, "States of Grace: Eureka Moments and the Recognition of the Unthought Known," *Journal of Analytical Psychology* 60 (1, 2015): 54–74.

44. Luigi Zoja, "Trauma and Abuse: The Development of a Cultural Complex in the History of Latin America," in Singer and Kimbles, eds., *The Cultural Complex*, pp. 78–89.

45. Dhar Prabhu, "Whispers in a Bull's Ear," p. 364.

46. Rudakova, "Diversity of Learning," p. 304.

47. Tsivinsky, "Impressions and Reflections," p. 637.

Chapter 4

Pioneers or Colonialism?

Henry Abramovitch

The Spread of Analytical Psychology

Analytical psychology, like psychoanalysis before it, was the creation of a single creative individual living in a sacred center. The spread of analytical psychology from this *axis mundi* occurred in a series of well-defined stages. The initial archetypal pathway for the diffusion of analytical psychology was as a sacred pilgrimage to the wise old man living in the holy mountain. Jung became the guru he claimed he did not want to be. Jung's original devotees resembled Christ's disciples who personally received the Holy Spirit through personal contact with the Master. To become a Jungian psychoanalyst, one needed direct contact with Jung or someone upon whom he had bestowed the grace of the "holy spirit." Systematic, or even previous professional training, was not required. Much more important was the understanding of the spirit of the work. These chosen disciples brought this new truth back home; or in the case of German Jewish refugees, like Erich Neumann, to new territories with an inspired passion while maintaining contact with the sacred center. In the subsequent phase, resembling that of Paul's Letters to new communities in Asia Minor, the main tension was between center-periphery in which these new semi-autonomous "churches" sought to establish their own identity and independence. Previously, analysts were self-selected and needed to have a "calling" rather than qualifications. The transition from charismatic solidarity toward institutional solidarity involved a new emphasis on professional training with formal rules for selection, training, qualifying exams, and other formal rites of passage. All societies were required to have codes of ethics as a communal guide for appropriate behavior rather than the prophetic inner voice of the Self.

Within this sociological perspective of the routinization of charisma, the phenomenon of Jungian analysts working in new territories may be seen as revitalization of an evangelical spirit to the international Jungian movement. Revitalization movements give renewed passion and vision to traditional, tribal religious groups, as Hasidism did to Judaism, or Ghost Dance did to Native American Plains Indians.[1] A new generation of missionary-analysts and supervisors set off to bring the word to those hungry for the "holy spirit." Ultimately, the project has not only created new and blossoming centers of analytical psychology, but it has also enriched the center.

Training at a Distance

This chapter will try to provide a personal view of training at a distance. It is based on in-depth interviews with visiting analysts and routers, my experiences doing lectures, seminars, workshops, supervising routers and Developing Groups members in various Eastern European countries, but especially doing ongoing group supervision in Warsaw and Moscow. I also did live supervision based on the topic of my lectures and presentations such as on siblings[2] and ethics.[3] I was one of the organizers of an intensive winter school in Jerusalem for routers from Kiev, Ukraine and Krasnodar, Russia in 2013, have done individual Skype supervision, and even supervised a router during her extended stay in Israel. My training contact with members of Developing Groups and routers has been unusually intense and diverse.

In some way, I had a distinct advantage. In addition to my psychological and analytic training, I am also an anthropologist. I am familiar with getting to know unfamiliar cultures or in anthropological language, to make the exotic, familiar and the familiar, exotic. I understood that I was engaged in intercultural communication and that I needed to learn from them so that they could learn from me. I also understood that if I was not to engage in unconscious imperialism, I had to know something of the ethos of these cultures and so I would read up about the culture, its fairy tales, literature, and the post-soviet dynamics. I also was experienced in working with translators and indeed it is a topic that I teach. I know that translation takes time, slows down the process, and simultaneously reveals and conceals. This is particularly true of culture-specific or "emic" concepts. There is a revealing joke about an American businessman who tells a long and complicated joke to a Japanese audience. The translator, however, says only two words and the entire audience bursts into laughter. Later, the American asked the translator how he managed to translate the joke so well. The translator replied, "It was simple. I said honored American has made joke. Please laugh!" The pragmatics of translation is often more important than the semantics.

Pioneers and Colonialism

Let me reflect on the two terms of the title: *pioneers* and *colonialism*. The word *pioneers* carries a positive emotional orientation that is in sympathy with the ethos of analysis. Pioneers do the work of opening up and exploring undiscovered territory. *Colonialism* carries a striking negative association today. It implies the deliberate and active imposition of political and economic rule on other peoples for their own good. In one sense, the distinction is one of time. Pioneers come first and colonials may follow. In terms of spreading the Jungian word, pioneers were those analysts who gave the first lectures, workshops, and seminars on analytical psychology to unorganized groups. They helped constitute the energy and vision needed to create what became Developing Groups and ultimately, the International Association for Analytical Psychology (IAAP) Router Program. The term *router* was coined by Murray Stein to guide those on the route to becoming analysts. It crystallized something of the essence of the process. The term provided the archetypal context of a route that was more than a journey, but less than a road map. Where pioneers have forged, bureaucracy must follow. Once groups were recognized and routers set on their way, colonials were needed as formal representatives of the IAAP in the form of liaisons, analysts, and supervisors working in a formal way with Developing Groups and especially with routers. In these terms, I was neither a pioneer nor a colonial officer but something in between. I was certainly involved in training, supervising, and teaching, but not in formal exams or assessments, although I often gave supervision against the background of the exam anxiety of my supervisees.

There is another difference between pioneers and colonials. It is how each defines itself in terms of the center-periphery. Pioneers set off to encounter the unknown, on their own, without a pre-established plan. Pioneers who enter and explore may often be in some danger. Subsequently, the identity of pioneers may be transformed. They can settle, move on, or even become a colonial agent. Colonials coming at a later point in time have a clear plan. The plan is to bring organization, values, religion, and profit. The colonials represent the Mother Country in the wilderness and must maintain standards of behavior. They are often subject to conflicts of loyalty between the locals and the authority far away, between those symbolically above them and those below. These conflicts can become acute as occurred in colonial America, or not, as in neighboring Canada. Clearly, there was an ongoing tension between how much the visiting analysts identified with the routers they were training.

It is important to consider how IAAP pioneers or colonials are perceived by native Developing Groups and routers. In general, pioneers are seen not as explorers but more as distinguished guests who must be received with honor. I discovered that each Eastern European country had its own style of honoring guests. In each country, there were intimate dinners. In Russia, there were formal ceremonies of farewell. Tourist trips often provided experiential insight into the collective and historical context of routers and their patients. In Moscow, a guided tour of Stalin's Cold War bunker provided an unforgettable experience into that collective nightmare. In Poland, the German occupation and the Holocaust never seemed far away. Georgia felt like a country moving simultaneously forward and backward in time.

The collective transference to pioneers is positive. The relationships with the colonial officers of the IAAP are necessarily more ambivalent. Developing Group members welcome the immense investment in their training but once formal training begins with structures and exams, a hierarchy is created within the group of those who are accepted on the route to becoming analysts and those who are not. The local cultural tradition of patron-client relation also plays a role. In Russia, for example, clients were expected to be dependent upon their patrons who were expected to look after them in a way very different from expectations between supervisors and supervisees in the West.

Being Chosen

How did I become involved in giving ongoing group supervision in both Moscow and Warsaw? It was not something I actually volunteered for, nor was I asked by the liaison. Rather, routers from each of these Developing Groups heard me speak on the topic "Illness in the Analyst" at an IAAP conference in Montreal.[4] In that talk, I tried to address this most difficult issue using a combination of active imagination, theoretical formulations, and my own personal and clinical experiences.[5] Based on that presentation, I was asked to come as a guest supervisor. I felt chosen. In doing supervision, I tried to continue something of the spirit of that presentation by combining what I call the "holy trinity" of theoretical formulation, clinical focus, and personal experience into an integrated supervising experience.

The Disciple's Dilemma

The first issue facing the visiting supervisor is what may be called the disciple's dilemma. Put simply, disciples represent the teachings of the master but must do so in a way that is authentic and individuated. Finding that balance between loyalty to the tradition and being true to one's Self is the disciple's dilemma. In clinical terms, the issue is when to be strict and when to be flexible. Working across culture requires clarity about the limits of flexibility. The key issue is to know when to show flexibility and when not. It forces visiting analysts to question their own tradition and to determine whether what is done is because it has always been done or because it has inherent value.

We all have a Jungian superego, the internalized collective voice of what Jungian analysis should be like. Being a disciple forced me to confront that superego more directly. It forced me to clarify my position on many substantive issues and to understand which were essential and which needed to be adapted to local practice. Catherine Crowther wrote that supervising in Russia "has required renewed thinking about my conventional boundaries, to declare what I feel is essential and irreducible about frame and what can be more flexible."[6] Crowther encountered striking differences in how the analytic frame was held. Russian therapists rarely saw patients four times a week, dealt with cancelations in a casual manner, and would set off for the long summer break without any return appointment. In addition to the striking differences in clinical norms, I felt that dissimilarities were rooted in cultural complexes that developed out of variances in how Christianity developed in Western and Eastern Europe, and specifically how Eastern and Western churches enacted the rite of confession. Jung claimed that the initial stage of psychotherapy involved confession. Catholic confession does resemble modern analysis in certain key aspects. Like psychoanalysis, it takes place in a strictly confidential frame at set times in which the priest is never seen directly but through an idealizing transference of the grille separating priest and parishioner. Except in extreme conditions, the closed confessional is the sole place where confession may take place. In contrast, confession in the orthodox tradition is done face-to-face, anywhere in the church. The believer is free to choose any spiritual mother or father (who may not even be a priest) to hear her confession. Eastern orthodox civilization assumes that confession (and so perhaps its descendant, analysis) takes place in a much more flexible frame.

I experienced another frame dilemma recently in Georgia. I was expected to hold individual supervision sessions in my rented accommodation which I had not yet even seen. My first impulse was to refuse, but then in a more analytic way, I decided to hold the tension until I could see whether my apartment could be used as a viable *temenos*. Ultimately, I discovered the apartment had a separate sitting room with a door that could be closed and which gave a sense of a clinical space good enough for supervision. The discussion with the group about my concerns stimulated their ethical awareness no less strongly than during my subsequent workshop on the same subject. I had also been warned about GMT—Georgian Maybe Time—but in fact, time frames were strictly adhered to. Since visiting analysts were in some sense acting as charismatic disciples representing analytical psychology, there was inevitably tension concerning their transference to Jungian theory and their tolerance of diversity or dissent. Training is always a high wire act. On the one hand, there is the need to respect and please one's supervisors and understand the signs both formal and informal they are sending. On the other hand, a student who adopts their supervisor's approach too diligently will be seen as a failure in the main task of individuating as an analyst. On the one hand, there is the expectation that trainees will develop competence, individuate, and become more and more themselves, while on the other, not displease those in authority who control the entrance into the promised land of being an analyst. This process is difficult enough in one's home culture but even more fraught when the foreign analyst carries the projection of Christ's disciples.

I know from speaking with supervisors and supervisees that power dynamics was a subject difficult to discuss. Visiting analysts or supervisors sometimes felt that "colonials" were merely going through the motions to fulfill the required hours of analysis or supervision as laid down by the IAAP. Once the hours were completed, even though the analyst felt the router was still in the midst of a learning process, the sessions ceased. The analyst felt used. Discussion with routers who had grown up under communist rule felt that analysts misunderstood the dynamics of a post-Soviet society. It was natural for them to give those in formal authority what they expected and dangerous to be too candid. The shadow of being exiled to a symbolic "gulag" for those who too openly challenged those in positions of power remained part of collective memory. Understanding the power dynamics of supervision may be one of the greatest challenges when working in different cultural settings.

Cultural Transferences toward the Visiting Supervisor

Doing transcultural analysis places an extra emotional and conceptual burden upon the analyst who must understand the interaction between the analysands' personal complexes and their collective, cultural complexes. To work effectively, the analyst needs to develop keen cultural self-awareness and an understanding of cultural transferences and countertransferences, i.e., how the analyst's cultural identity is perceived by their analysands and vice versa. Similar dynamics are played out in group supervision.

Doing group supervision away from home, I experienced intense expectations in the form of a special, performance-persona anxiety. I had high expectations of myself. Supervision at home is an unfolding process which takes place at its own pace. Not every session needs to be dramatic. Coming from far away and carrying group projections, I felt an adrenaline-fed impulse to supervise like one who can transubstantiate the "bread and wine" of clinical routine into the "blood and body" of the Self.

At the same time, I sensed the danger of transference from routers toward me as their supervisor as an infallible wise old man. As one observer of supervision, Levenson has remarked supervisees often want certainty:

> This method of teaching operates on a manual of prescribed situations and responses. It is also inevitably lock-stepped into a metapsychology which is both authoritarian and omniscient. For everything the patient says and the supervisee does, the supervisor has a theoretical formulation and a corresponding piece of behavior. The patient is explained to the supervisee in terms of the metapsychology, and the therapeutic intervention follows automatically, to wit: 'The patient is narcissistic and unable to ... and therefore one must ...' I confess, I find this process as abhorrent as painting by the numbers, and I think the outcome is about as predictable and esthetically miserable. This neat fit between theory and practice is, I believe, the last resort of the unimaginative; but it has considerable appeal and tends to make the supervisor rather popular among psychoanalytic candidates who wish to believe that someone is clear about what they are doing.[7]

As a supervisor, I had to hold the tension between allowing these group projections but not identifying with them, and between knowing and not knowing.

In addition, there were anticipatory transferences to me as a supervisor and the role I was expected to play. One supervisee had read Jung intensively but in a rather idiosyncratic fashion. He was very interested in topics such as alchemy and astrology, which admittedly were very important to Jung. Although I do have colleagues who value alchemy and even astrology in their clinical work, neither symbolic system is something I use in my clinical work. I sensed the shock of this pioneering reader when he discovered he wanted a Jungian psychology I was unable to provide. It was as if the analytical psychology he believed in was not the one I had come to teach. This disappointment provided a valuable opportunity to discuss with him and others their images of what a "real" Jungian analyst was.

Working across Culture

Jung stressed clearly the importance of seeing a person in his own context and culture to understand symptoms, dreams, and key symbols. As Higuchi, a Japanese analyst has written, "If you receive a person, you also receive his entire society."[8] The dilemma for the visiting analyst is how to receive supervisees together with their entire society. Treating or supervising someone from one's own home culture involves a certain cultural coziness; it is assumed that analyst and patient will understand the unspoken cultural codes and references as well as key metaphors. For example, if a patient mentions The Ashes, every British person will pick up the reference. If an Israeli says, "I am not your friar," few non-Israelis would understand the aggressive reference to a cultural complex involving a profound fear of victimization; or if a religious Jew says in response to an intense session, "I felt like Rabbi Shimon Bar Yochai"— whom Jung encountered in his post-heart attack Visions[9] —the patient is referring to Rabbi Shimon's experience upon coming out of the cave after years of mystical illumination. How many of the uninitiated would understand the reference to intense, burning rage at the banality of the mundane following an intense transformative experience? Archetypal experiences, such as mothering, befriending, or feeling betrayed, for example, are always coded within a specific cultural and historical context.

British society's shared cultural values often transfer into analysis, such as affective restraint, self-reliance, the importance of being on time, and speaking honestly about one's feelings about family members and the analytic relationship. In many cultures, these assumptions would not hold. Working with unfamiliar others, analysts often encounter the *unheimlich* (uncanny, literally un-homelike) sense that normal, expected conditions for analysis are absent or missing.

Words That Cannot Be Translated

Analysts face many issues working across cultures, but I will focus only on the cultural use of language, the use of translators, and what can become lost in translation. Anthropologists have pioneered a number of techniques for getting at the heart of a culture. One of these is to explore keywords that embody central signifiers of the culture as a whole and which require thick description to understand. Sapir-

Whorf's hypothesis even suggested that the nature of language affects the nature of thought.[10] This controversial hypothesis has received little attention from the psychoanalytic community; however, Wittgenstein remarked that the limits of my languages are the limits of my world.[11] Differences in semantic and affective fields undoubtedly play a role in how we perceive the world. The difficulty is that speakers, whether analysts or supervisors, do not recognize the limitations of their own languages. Often, an emotional gap is created when there is no corresponding word in the other language. Even with the best of translations (e.g., translators trained in Jungian psychology, or bilingual routers who act as translators), visiting analysts have serious difficulty comprehending untranslatable terms and their cultural associations.

It may surprise the reader that Hebrew has no word for *mind* or *solitude*, nor does the language make a distinction between *envy* and *jealousy*, or even between *like* and *love*. On the other hand, there were many times in supervision when I felt the need to use Israeli terms that have no equivalent in other languages. One example is *dugri*, which is usually translated as "straight-talk." It refers to an Israeli cultural mode of speech saying what you really think without consideration of the other person's feelings. It is a characteristic of Israel's cultural complex and assumes that relationships are robust and can withstand conflict.[12] In many cultures, such straight-talk is considered rude. At those times in supervision, I would teach the group the meaning of *dugri* and then might ask the struggling supervisee, "please, *dugri*, what do you mean to say?" It allowed the supervision specific directness in cultures that favor indirectness.

An opposing danger is that the analytic visitor does not comprehend what is missing in the translation. Translations reveal as much as they conceal. To understand what is missing requires a pioneering spirit. For example, a Russian supervisee says that a patient tells her that she feels like a friend towards her. As my mind starts to think about the transference implications, I wonder about the Russian word she used. She does not say friend, but *drug*. The only possible translation for *drug* is friend but this is certainly not an adequate translation. *Drug* and friend occupy very different emotional and interpersonal space. The English word *friend* covers a vast semantic range from *best friend* to a *recent acquaintance*. It is supremely ambiguous concerning the nature of the social and emotional relationship and any mutual obligations involved. The Russian language has a rich vocabulary for the type and depth of friendship that has no equivalent in the English language.[13] In Russian, *drug*, indicates someone very close and implies an intense, demanding, enduring relationship that is both exhilarating and exhausting. *Drug* is a deep friend who can ease the pain of life and is more of a soul brother or soulful sister. In Soviet times, your *drug* was the one person who could be chosen freely; for only children, they took the place of missing brothers and sisters, with whom there was a total commitment. There is another term, *podruga*, which is less than *drug*, but still stronger than what is usually intended as *friend*. Polish, too, has six terms for degrees and genders of friendship with the most important, *przyjaciel*, indicating a strong loyalty and attachment bordering on love. Not understanding that Russians and Poles have specific words for relationships with more depth, devotion, and duration than is implied by the English word *friend* cannot be without consequences for analysis. When I listen to those cases involving a positive transference to the therapist, I sense there is a culture-specific *drug*-like or *przyjaciel*-like relationship that primes patients to open their souls to their therapists. In this sense, there is a cultural concept or positive cultural complex that connects to the therapeutic situation that does not exist in the English-speaking world.

Studying such keywords as *drug* or *przyjaciel* is one important technique that anthropologists use to understand culture in depth. What is true for concepts like *friendship* is even truer for words used to describe inner states so important for the word of analysis. The Russian language seems blessed with such words that defy translation and, for example, express human defects and existential emotional states in a single, pitiless world. Consider the term, *poshlost* for which none of the list of English words (*cheap, sham, common, smutty, pink-and-blue, high falutin, in bad taste*) are adequate. No English word can convey what is implied by *poshlost*, namely, an attempt at unmasking, exposing, or denouncing cheapness of all kinds. The Russian word *poshlost* is so cleverly covered over with protective tints that its presence often escapes detection. In other words, it reflects and documents an intense awareness of the existence of false values and the need to deflate them. And all that in one, simple, word. Other untranslatable terms mark out a fascinating if unfamiliar emotional-moral-semantic territory, such as, *dusha*, usually translated as *soul*, but actually it is the inner space in which a person's emotional, spiritual, and moral life, and presumably analysis, goes on.

One further word I discovered was *toska*. No single word in the English language renders all the shades of toska.

> At its deepest and most painful, it is a sensation of great spiritual anguish, often without any specific cause. At less morbid levels it is a dull ache of the soul, a longing with nothing to long for, a sick pining, a vague restlessness, mental throes, yearning. In particular cases it may be the desire for somebody or something specific, nostalgia, love-sickness. At the lowest level it grades into ennui, boredom.[14]

It is a cousin of the Portuguese *saudade*, a nostalgia for something you have never known that is the basis of the emotive vocal tradition of Fado. *Toska* seems to embody something uniquely Russian and yet universal. It gives me insight into the psychic and spiritual suffering which is so common when it is named. When I can say in supervision, "Does your patient experience *toska?*," I get surprised but gratifying looks of how an outsider could understand this insider's term. The group moves closer. The unusual thing for me was that *toska* is seen as an existential state and *not* something that needs to be explained developmentally.

Another word I learned was *nedoperee'pel*, a drinking-related word meaning "under-over-drunk." This special term referred to someone who drank more than he or she should have, but less than he or she could (or wanted to) have. Slavic languages are gifted in describing the breadth and depth of alcoholic space, while Hebrew is not. Many of the cases I supervised had alcoholic behavior deeply embedded in them often in the archetypal form of an alcoholic, abusing, and abandoning father. It seemed to have a predetermined fate-like aspect. Here I was called upon to supervise such cases whereas in my practice in Israel, I had limited experience. As in any good supervisory process, I was learning from my supervisees as they were learning from me. Sensitizing oneself to such keywords, I believe, is a helpful approach for visiting analysts to avoid linguistic colonialism and to enter into the collective unconscious of the people.

The Hardest Experiences

What were the hardest experiences? Normally, though not always, supervision was done with an interpreter. This required a somewhat structured and unnatural progression where only one person would speak at a time. Occasionally though, in the middle of a case, everyone in the group would begin speaking rapidly together in their native tongues leaving me linguistically ostracized. Ultimately, some order would return and the interpreter, often with help of group members, would try to translate what had been said and why people had become so excited. Sometimes, the reversion to native tongue reflected tensions within the group, or within the Developing Group, or broader political tensions, and those were much more difficult to translate or understand. In some groups, internal conflicts and splits would interrupt the training process. Normally, as a group supervisor working on home ground, I could use my inner space to sense what was going on in the group and then decide to intervene appropriately or not to intervene at all. In this case, I felt off balance, not sure of whether what I was feeling was connected to the unconscious life of the group.

Finally, there are delicate dilemmas concerning the transition of a router to the status of analyst, and collectively for Developing Groups to the status of Society and then Training Society. Here, the situation resembles the struggle for autonomy between a parent and adolescent. Analyst-parents may be concerned when is the right time for the adolescent-router to go solo; while routers, in turn, may experience the analyst as an overbearing (or in other cases, abandoning) parent. Here, the collective transference seems much more mixed with feelings of liberation intermingling with a sense of abandonment.

Endings

During my visits, I would try to stay at the same hotel. In retrospect, I realize I was seeking an island of continuity and sense of home as a stranger in a strange land. There were cultural differences in farewell. My visits to Moscow typically concluded with a farewell ceremony involving speeches, gifts, and many hugs. Endings in different places were all so unique that it made me reflect about how deeply culturally embedded the nature of endings can be. I think of the Russian tradition of sitting in silence before a journey. It made me contemplate my own endings, not only when I left the foreign lands, but later when I returned home.

Throughout the process of training at a distance, the role of the explicit and implicit image of the analyst plays a crucial part within the training process, as implied in the following question: How do you know a Jungian analyst when you see one, when the very image of the analyst is itself changing?

Pioneers or Colonialism? Or Both?

To return to the question of whether visiting analysts were pioneers or colonials, the answer, of course, is that we were both. We opened up new territory and then gave it structure as Developing Groups, Router Programs, institutes, and training institutes. We spread "the word" and created "disciples" and these "churches" flourish. Bringing the "gospel" to the periphery at the same time has revived the Center with new creative energy. The heartland of analytical psychology may no longer be in the sacred mountains of the West, but increasingly flourishing in China and many countries of the former Soviet Bloc. One reason why analytical psychology is so strongly received in these societies is because Jungian psychology uniquely provides a conceptualization of the dynamic tension between the collective and the individual that is both conscious and unconscious. It is a tension that reflects the daily life of members of these societies where collective pressures are so great on patients and routers alike. Under Soviet rule, analysis was not only banned, but there was no private space in which to conduct it.

New analysts from the former Soviet countries have much to teach us about the dialectic of the collective and the personal. Some of the most innovative, recent papers are contributions from post-router analysts. These essays include Elena Pourtova's writing on the clinical importance of nostalgia,[15] the article by Malgorzata Kalinowska on cultural complexes and collective memory in Poland,[16] and Elena Bortuleva's article, "Rivers of Milk and Honey—An Exploration of Nurturing the Self in a Russian Context."[17] Bortuleva discusses the words for nutrition (*pitanije*) and education (*vos-pitanije*) that are identical except for the prefix *vos-*. The prefix adds to the word the idea of a gradual addition to the core of all that makes a person into a person, transcending from actual into symbolic, and helping a person "grow spoon by spoon" giving "nurturance at the wider human, or cultural level." The caregiver who remains sensitive to the ever-changing needs of a child, starting with the baby's feeding preferences, can give the baby the fullness it requires, drop by drop, to transform *pitanije* (nurturing with food) into *vos-pitanije* (nurturing with truly human experiences). Using these culture- and language-specific ideas, Bortuleva is able to give, via a clinical illustration, new insights into how the primary self grows and develops into a mature Self. The contributions of Pourtova, Kalinowska, and Bortuleva are only a few examples of how new teachings from what was once the periphery are flowing back to enrich that Center.

Conclusion

Jung was a pioneer of the unconscious, but he was a reluctant colonial. He was deeply suspicious of the de-individualizing dangers of organizations and the imposition of values from without. He originally opposed the formation of an institute in Zürich that would carry his name, and only reluctantly consented to the formation of the IAAP on the occasion of his eightieth birthday.[18] Yet, Jung worked hard to spread his teachings. He travelled. He lectured. He gave seminars and workshops. He was a charismatic leader who had a truth he wanted to spread.

How Would He Feel about the Developing Groups and Router Programs?

It is likely that he would see them as a natural continuation of the psychology clubs that also began in Zürich in 1916 and spread to many places. He would likely approve that Developing Groups come up from the bottom, in a sense that they must spring from an inner need and only subsequently come under the auspices of the IAAP. The necessary bureaucratization of training would most likely encounter his disapproval. In this he may resemble Moses who wished that the entire people be prophets, but this was impossible. Jung was a great pioneer of "the way." We, who follow after him, strive to be colonials who are also still pioneers, each in our own way, and of that, Jung surely would have approved.

Notes

1. Anthony Wallace, "Revitalization Movements," *American Anthropol* 58 (1956): 264–81.

2. See especially chap. 7 in Henry Abramovitch, *Brothers and Sisters: Myth and Reality* (College Station, TX: Texas A&M University Press, 2014), pp. 87–107.

3. Henry Abramovitch, "Stimulating Ethical Awareness during Training," *Journal of Analytical Psychology* 52 (2007): 449–61.

4. Henry Abramovitch, "Illness in the Analyst," in Pramila Bennett, ed., *Facing Multiplicity: Psyche, Nature, Culture. Proceedings of the XVIII*th *Congress of IAAP* (Einsiedeln: Daimon Verlag, 2010).

5. Henry Abramovitch, "Into the Marginal Zone," in Patricia Damery and Naomi Lowinsky, eds., *Marked by Fire: Personal Stories of the Jungian Way* (Carmel, CA: Fisher King Press, 2012), pp. 94–109.

6. Catherine Crowther, "Supervision of the Apprentice," in Murray Stein, ed., *Jungian Psychoanalysis: Working in the Spirit of C. G. Jung* (Chicago & LaSalle, IL: Open Court, 2009), p. 395.

7. Edgar Levenson, "Follow the Fox—An Inquiry Into the Vicissitudes of Psychoanalytic Supervision," *Contemporary Psychoanalysis* 18 (1982): 6.

8. Kazuhiko Higuchi, "Jungian Psychoanalysis in the Context of Japanese Culture," in Stein, ed., *Jungian Psychoanalysis*, p. 246.

9. C. G. Jung, "Memories, Dreams, Reflections" (New York: Random House, 1961), pp. 325–28.

10. Paul Kay and Willett Kempton, "What is the Sapir-Whorf Hypothesis?," *American Anthropologist* 86 (1, 1984): 65–79.

11. Ludwig Wittgenstein, *Tractatus Logico-Philosophicus*, trans. D. F. Pears and B. F. McGuinness (London: RKP, 1961), Prop. 5.6.

12. Tamar Katriel, *Talking Straight: Dugri Speech in Israeli Sabra Culture* (New York: Cambridge University Press, 1986).

13. Natalia Gogolitsyna, *93 Untranslatable Russian Words* (Montpelier, VT: Russian Life Books, 2008).

14. Vladimir Nobokov, "Commentary," in *Eugene Onegin: A Novel in Verse by A. S. Pushkin* (Princeton, NJ: Princeton University Press, 1991), p. 141 and n. 8.

15. Elena Pourtova, "Nostalgia and Lost Identity," Journal of Analytical Psychology 58 (1, 2013): 34–51.

16. Malgorzata Kalinowska, "Monuments of Memory: Defensive Mechanisms of the Collective Psyche and Their Manifestation in the Memorialization Process," *Journal of Analytical Psychology* 57 (4, 2014): 425–44.

17. Elena Bortuleva, "Rivers of Milk and Honey—An Exploration of Nurturing the Self in a Russian Context," *Journal of Analytical Psychology* 59 (4, 2014): 531–34.

18. Thomas B. Kirsch, *The Jungian: A Comparative and Historical Perspective* (London and Philadelphia: Routledge, 2000).

Chapter 5

Cultural Complexes and Working Partnerships

Catherine Kaplinsky

> Wars, dynasties, social upheavals, conquests, and religions are but the superficial symptoms of a secret psychic attitude unknown even to the individual himself, and transmitted by no historian.
> —C. G. Jung, "Civilization in Transition," § 315

Introduction

We are living in times of major social upheaval—what Bauman calls "liquid times." Sweeping technological change, faster travel and communication, global advertising, and infiltration of multi-national corporations have all created problems of conflicting values and inequality while also creating possibilities for transformation.[1] Our cultural differences are being both flattened out and brought into sharper focus.

One new possibility that has been opened up is that we can now take analytical psychology to various parts of the world. In Jung's time, this would have been unthinkable. Jung was clearly fascinated by the influence of culture and history on the individual psyche and he brought many profound insights to this topic, but he often seemed to lose sight of the fact that his own attitudes and theories were, themselves, a product of his particular time and place.[2] Learning from him, we must be conscious of our own attitudes, our cultural histories, and the effect they have had on our theory and practice.

Inevitably, when working abroad as visitors, we are drawn into the collective unconscious of the culture in which we are guests and at times this can leave us feeling bewildered and at odds, especially in the transference and countertransference dynamics. Our way of working simply may not fit with the values of the culture. It is important that we remain conscious of our limitations, careful of the dangers of hegemony while trying to stay true to the principles of our profession. This is both demanding and humbling. Essentially, we are on a quest for creative solutions, open to the realities of the culture in which we are newly engaged in partnership and open to learning from them.

Significantly, Jung did not propose a particular intermediate realm between the personal and the collective. It was H. G. Baynes who first postulated the idea of a "cultural psyche" in 1939 and, subsequently, Henderson expanded our understanding of the cultural unconscious.[3] Henderson defined it as

> an area of historical memory that lies between the collective unconscious and the manifest pattern of the culture ... it has a kind of identity arising from the archetypes of the collective unconscious, which assist in the formation of myth and ritual and also promotes the process of development in individuals.[4]

He pointed out that much of what Jung called "personal" was actually culturally conditioned. Later, Adams says that "much of what Jung called 'collective' was actually 'cultural.'"[5] So here we can already see the emergence of dialectical interplay.

Essentially, the cultural unconscious is a dynamic field, a living history, connecting simultaneously both to the collective and the personal. Culture influences the way individuation processes are lived out and in turn the processes of individuation influence the culture. Identity and culture are always entangled. Figes writes,

> Identity is not meaningful unless one can show how it manifests itself in social interaction and behaviour. A culture is made up not simply of works of art, or literary discourses, but of unwritten codes, signs and symbols, rituals and gestures, and common attitudes that fix the public meaning of these works and organize the inner life of a society.[6]

Dawkins introduced the term *meme* as the cultural equivalent of a gene.[7] He sees memes as the replicators of cultural information that one mind transmits to another. There has been much debate, however, about the validity of this concept. McGilchrist argues that if they existed, they "would be replicating, unlike genes, within a mind: a mind whose constant interaction with whatever comes to it leaves nothing unchanged … We are imitators not copying machines."[8]

Jacobi in "*The Psychology of Jung*," produced a diagram which illustrates the relationship between the area of archetypal collective unconscious to culture and the individual.[9]

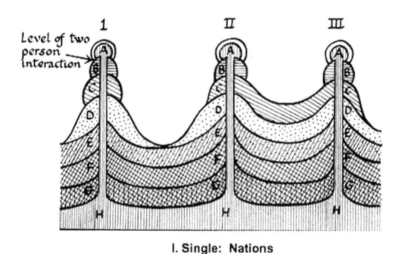

I. Single: Nations

II and III: Groups of Nations

A. Individual   E. Groups of Nations
B. Family   F. Primitive Human Ancestors
C. Tribe   G. Animal Ancestors
D. Nation   H. Central Force

Figure 1. Jacobi's "psychic family tree." [Adapted from Jacobi's Diag. 11 (1942), p. 34]

Today, Jacobi's diagram might be less well-defined due to the "liquid times" in which we live. As Bauman has emphasized, the structures, institutions, routines, and patterns of behaviour are becoming increasingly less reliable and the boundaries which help to define us, less secure.[10] We are even losing our familiar climate. These are the insecurities of which we are all aware as we reach out towards other cultures. Significantly, this "reaching outwards" is itself part of an unconscious intercourse with the capacity both to disturb and enrich.

CULTURAL COMPLEX, PERSONAL COMPLEXES, AND REFERENCE GROUPS

Singer and Kimbles developed the idea of the cultural complex, which is helpful when exploring this energetic field.[11] It facilitates the exploration of both the cultural unconscious, with its impact on groups and institutions on the one hand, and individual psychology on the other. It also enables us to think about different cultures in a fresh way. The idea of the cultural complex is clearly derived from a coupling of Jung's theory of complexes with Henderson's theory of cultural unconscious.[12]

To recap briefly on the theory of the personal complex, it was Jung's work on the Word Association Test that led to his theory of complexes. He also called them "splinter psyches" and then again, the "architect of dreams."[13] The complex is a collection of images and ideas clustered around a core of one or more archetypes and is crucially characterized by a particular feeling tone. Complexes are rooted in the body and when activated affect our behaviour. They are natural phenomena, essential to psychic life and can develop along positive or negative lines. They can therefore involve a certain turning back in on oneself—either a recoiling, or an excited reaching-out.

As analysts, we often discover complexes through transference and countertransference dynamics with our patients and we can think of them as energies and feelings "trapped" in the body and on the ego-self axis. Our task is to try to reach into these intangible transitional areas and join with them as the dynamics are replayed. This experience also tells us something about the history of our patients and the development of dammed-up energies. Inevitably, the feeling tones spiral out into the world of family and society and hint at cultural patterns. Shifts may occur as a result of the newly experienced affect.

Cultural complexes are elemental aspects of any culture. In Singer and Kimbles' words, they can be thought of as arising out of the cultural unconscious as it interacts with both the archetypal and personal realms of the psyche [as well as] the broader outer world arena of schools, communities, media, and all other forms of cultural and group life. As such, cultural complexes can be thought of as forming the essential components of an inner sociology.

This inner sociology is an essential aspect of the cultural complex and addresses the dynamic relationships "of groups and classes of people as filtered through the psyches of generations of ancestors [and] has all sorts of information and misinformation about the structures of societies."[14]

Singer and Kimbles emphasize that "[o]n an individual level, cultural complexes are expressions of a need to belong and have a valued identity within the context of a *specific reference group*." The criteria required to belong, however, might lead to splitting, rigidity, and a whole range of phenomena that we recognize as psychological disturbance: "At the level of [the] group, cultural complexes seem to offer cohesion which provides a sense of kinship and group spirit. [But], at the pathological extreme, this kinship is expressed in archetypal defences of the group spirit."[15]

It is this "need to belong" and the "need for a valued identity within the context of a specific reference group" which I would like to focus on further. To belong to a reference group implies a particular "fit," a similarity of identity and clearly harks back to early attachment dynamics. Aspects of ourselves that do not "fit" must therefore be relegated, split-off, or repressed. This can result in splinter psyches or complexes. Redfearn explores this in his concept of sub-personalities.[16]

Shadow and Archetypal Morality

A discussion of the cultural unconscious must include Jung's concept of the shadow. A simple definition of the shadow is "the thing a person has no wish to be."[17] This links to "what is acceptable" and "what is not acceptable." The personal shadow, as we know, is more easily accessible and relates to the repressed and disowned part of the individual. It is acquired as an adaptation by the ego to outside demands. So, "that which a person does not wish to be" is moulded by what mother and intimate others wish him or her to be, as well as by the ideology of the various groups which make up the interactive field of the culture in which they are contained.

With the concept of the cultural unconscious comes the cultural shadow. Stevens expanded on this idea in his book *Archetype*.[18] Like the personal shadow, the cultural shadow arises out of the archetypal area, an area of non-ego. From this emerges ego development, me/not-me dynamics, and good and bad values which extend out into the world. All cultures advocate good values and shun bad values and this lends support to the idea of a kind of programming with a phylogenetic, or archetypal basis, for morality.

The moral complex, or Freud's super-ego, "is the psychic organ that makes society possible." It monitors our behaviour so that adherence to cultural values is ensured. Non-conformance to the moral code brings threat of abandonment, ostracism, or other forms of punishment. But there is a price and paradox to morality since, to quote Stevens, "the very milieu that makes actualization of the Self possible also demands that aspects of the Self remain un-actualized in the unconscious or be actively repressed."[19] Herein lies the cultural shadow.

While different cultures have different ways of actualizing the moral code, different aspects of the Self are repressed. From anthropology we know that what is defined as "good" and "bad" is highly variable culturally, making the idea of universal values problematic. For instance, different cultures demand that behaviours such as adultery, homosexuality, incest, and overt expression of aggression, are suppressed to a greater or lesser degree, so that which is required to be repressed will vary from person to person and culture to culture. What does seem universal, however, are these particular nodal points around which all cultures develop moral codes.

There are some overlapping themes between Stevens' work on the cultural shadow and Singer and Kimbles' cultural complex. Both address an inherited archetypal form or pattern of morality. These are then fleshed out in moral codes which vary according to environmental and historical change. It follows that cultural complexes are particular expressions of an archetypal morality and also that individuation processes are shaped to some extent by such complexes.

Within this complicated dynamic web, born out of archetypal longings to attach, belong, and be validated, are the manifold transitional spaces in which integration takes place or in which splits and rigidities occur. These are the spaces where boundaries evolve and interact and in which personal and cultural complexes evolve. This is the stuff of our work within the analytic *vas* where these themes are played out in the transference and countertransference and in regressive states. The search for meaning circulates around a longing for a state of "in-ness," which in turn involves a sense of agency and of being an effective part of life's drama and process. This search spirals out into the culture for each individual.

A Bedouin saying quoted by Chatwin illustrates the spiral-like nature of our sense of "in-ness and belonging."[20] It reads,

> I against my brother
> I and my brother against our cousin
> I, my brother and our cousin against the neighbours
> All of us against the foreigner.

The requirement of otherness in order to create a sense of boundary and identity is well evidenced from infant research and we know too that while boundaries provide a sense of containment, they can also be an area of exchange where opposites can meet, or they can be a point to be broken through or out of, in order to "become" and individuate. Boundaries also shift and change; they defend, can become rigid, and can develop into barriers. Shadow projections are contained within boundaries keeping out the unacceptable. They can also become pathologically rigid and are a source of conflict and war. As in the Bedouin saying, they can spiral out either into benign cultures or ruthless regimes. Essentially, at an individual level, processes of rejection and projection and boundary-making are expressions of an archetypal push to separate out and individuate. Boundaries, containers, and a sense of belonging then spiral out into the world of us/them dynamics and of righteousness. Essential to this process is a place to locate our aggressive energies. This is explored in detail in Gordon-Montagnon's paper "Do Be My Enemy for Friendship Sake."[21]

Infant Research: Ego Boundaries and Skin

Recent infant research has expanded our understanding of these early processes. The ego evolves within transitional boundary areas where unconscious choices are continually being made. These are choices about what must be taken in and accepted and what is rejected or projected. Fordham expanded our understanding of the early dynamics involved in separating out and the development of ego boundary.[22] Crucial to the process, is the archetypal energy involved in reaching out with curiosity (what Fordham called *deintegration*), followed by the processing of these experiences (what Fordham called *re-integration*). The rhythm of deintegration and re-integration can be understood as a primary expression of the individuation principle. There is a bodily equivalent too in the immune system where mechanisms within an organism protect against infection by identifying and killing pathogens. Inevitably, there can be complications when the immune system can become disordered and research links have been made between emotional states and some immune deficiency disorders, e.g., rheumatoid arthritis and Graves' disease.

The experience of skin is particularly important in relation to early ego boundary. Freud was the first to write about this. He wrote in a footnote that "the ego is ultimately derived from bodily sensations, chiefly from those springing from the surface of the body. The ego is thus a mental projection of the surface of the body."[23] Bick was the first to conduct an in-depth study of the psychological function of the skin, postulating that the skin is experienced as a sheath that holds together the psyche and soma.[24] Through the experience of adequate holding and shared satisfaction, the infant is able to tolerate periods of separateness which then create spaces in which develop the imaginal and the capacity to symbolize. Since Bick's research, much work has been conducted illuminating the relational significance of the skin to ego boundary. Anzieu speaks of an "ego skin"; Feldman of "psychic skin."[25] Essentially, the "skin ... provides the point of contact with the external world ... [and] acts as a delineator of boundaries between what is experienced to be outside and what is experienced to be inside the self."[26] The colour of the skin is of course irrelevant to the infant other than signifying familiarity.

Difference and otherness must first be established before there can be dynamic relating and the consequent possibility of union and harmony. Neuroscience provides interesting evidence in this area. Crucial to this process is our understanding of mirror neurons and the role they play in imitation. When we imitate, it is as if we are experiencing that which we imitate and when this happens, the firing of neurons is replicated. Then, with the development of imagination when there is no visual or other stimulus, it transpires that some of the same neurons are activated again—as if there is a re-experiencing of that which was imitated but now imagined.[27] Attachments and bonds develop with those whom we imitate, beginning first with mother and then spiralling out into family, groups, and culture beyond. Here lie the seeds for dynamics to both compete and cooperate and may be considered too to underlie the potential for either war or peace.

In order to understand cultural complexes, I have explored a spiralling, complicated but dynamic web developed in the service of boundary-making. This dialectic dynamic reaches out from early cell division, through ego development, family, groups, institutions, into the culture, and back again. Crucial to this process of boundary-making and becoming is imitation of the other, mirroring, and feedback. Splits, rigidities, and complexes inevitably develop along the way in relation to differing personal and social histories. Our job as analysts is to try to reach the secret psychic attitudes Jung writes about in the quote at the beginning of this chapter. We must home-in and join with them in order to create space for creative shifts. These are the transitional areas in which the new/the other are allowed in and integrated. With adequate containment and free expression of the self, boundaries can soften and become more flexible, bringing about changes of attitude. Homing-in and joining with those from different cultures inevitably brings both enrichment and challenges.

Working in Different Cultural Settings

I will discuss three different examples illustrating cultural complexes based on my experience working in different cultures. These come from my own observations, inevitably influenced by my own cultural experiences including shadow aspects. It is important to note, also, that the very process of observing plays into the ever-shifting interactive dynamic.

Russia

Russia is one of the countries with whom the International Association for Analytical Psychology (IAAP) has been able to establish a fruitful, mutual working relationship aimed at reviving analytical psychology after a period in their oppressive history when it had been forbidden. This collaboration began in St. Petersburg in 1998, expanded to Moscow in 2000, and later to Krasnodar, in the south of Russia and Kemerovo, in Siberia. In St. Petersburg and Moscow more than twenty UK analysts from all four London-based Jungian Societies would visit regularly—some to teach and supervise, others to give shuttle personal analysis. The work was usually with interpreters present and the aim was to help Russian trainees of the IAAP Router Programs with an interest in Jungian analysis to gain an international qualification and ultimately to found their own Society and later Training Society.

My role as supervisor in Moscow was a privilege and life-changing. Arriving at our small hotel late on a Friday afternoon and starting work an hour later in cellar-like rooms in the Institute opposite felt like a brutal dive into an emotional world which embraced, rejected, fascinated, and was always very intense. I came to think of this particular intensity as a symptom of Russia's history which preceded the upheavals of the twentieth century. Coming from the West, with knowledge of the Great Russian writers, we appreciated what seemed like a natural gift to "go there." Freud certainly recognized this gift and seemed to think Russians were closer to the unconscious than Western people.[28] Quickly, there was an ability to reach emotional depths which had been lying low and there was a very real appreciation of the space provided, the *vas*, in which together we could "be" and think about emotional issues.

It would be impossible to attribute a single cultural complex to Russia, the largest country in the world, spanning two continents—Europe and Asia—and it might seem arrogant to do so; however, dipping into the culture provided glimpses of the complex issues arising from their more recent history of forbidden self-expression, of ruthless oppression, of gulags, of torn relationships—a history in which processes of individuation were stifled. We learned very soon that our traditional way of working was going to be challenged. First of all, the supervisory rhythm was different to that which we were used to in the UK. Instead of weekly or fortnightly, each visiting supervisor offered four very intense weekends a year in which we supervised people both in groups and individually. The translator created a powerful "third" on whom we relied for accurate accounting and impartiality.

Early into the work, we learned about key differences between our cultures. For example, a suggestion to write a letter to a client—if, for instance, they had not turned up for sessions—would result in raised eyebrows and a furtive, exasperated exchange of glances. Texts were considered the only reliable, "trusted" means of communicating we were told. We learned too that it was "polite" when talking on the phone to know who was speaking without asking for a name, and later we learned that it was common for the trainees we were supervising not to know the full name of the clients they were working with, though we were informed that things were changing and one trainee suggested that not asking the full name of a client could be a sign of the trainee's personal paranoia. It was clear that everyone knew to suspect intrusive surveillance. In time, we also learned about "the kitchen ego" and "the street ego." The gossip and the "real" exchanges happened in kitchens and passages. Inevitably, we were all part of the gossip and mistrust. We were each given nicknames, we discovered, just as those in power in Russia had been given nicknames; for example, *The Red Tsar*: Stalin, or *Money-Box*: Roman Abramovich. We had to earn trust. As guests and newcomers, we had to respect the culture while at the same time keeping in mind the ideal authentic exchange. The "street ego" on the other hand was for "formal" presentations, acceptable to the powers that be. We knew to be aware that we might represent an intrusive "power." The "street ego" had arisen out of fear. Fear and control can result in the denial of self-expression so that unexpressed libidinous energy becomes knotted in the body and leads to the formation of complexes with particular feeling-tones. These, in turn spiral out into the culture in which they were born. Part of the Russian cultural complex then is around a fear of being authentic.

The *matryoshka*, the Russian nesting doll, is a symbol of a related aspect of Russian culture though it was only "born" in the 1890s during Tsarist Russia. It is a particularly relevant symbol for authentic emotional growth and fits well with theories of infant research, especially around the idea of an appropriate "holding" space which allows for creative growth and individuation. As we know, when the space between is too great, a sense of abandonment can lead to a rupture in the relationship between the ego and the self. When the space is claustrophobic or intrusive, there is also a rupture but the feeling-tone would be different.[29] The *matryoshka* is a symbol of the ideal holding space as development proceeds. Ledermann has written about Russian dolls in relation to narcissistic disorders, particularly when some of the inner dolls are missing leaving empty space.[30] Bortuleva thinks about them in relation to a Russian Cultural complex.[31] Over the years the "doll" emerged frequently in cases brought by supervisees. A doll can be played with, dressed-up, and discarded. Female clients were often described as "porcelain dolls" or "Barbie dolls" and the implication was of an inner emptiness and a (false?) persona. Often the falseness was literal with cosmetic "work" having been done on many parts of the body in an attempt to become more acceptable. The inner emptiness usually had a history of traumatic rifts and emotional upheavals represented by an empty *matryoshka* or with some missing inner, baby dolls. There were stories of gulags, of moves to Siberia or elsewhere, or being sent off to be looked after by grandmother, and back to a mother, no longer recognized. There were stories of fathers or grandfathers who had been part of the old Soviet regime, now displaced and humiliated, or a background linking to the old Tsarist royalty or oligarchs with stories of bullying and abuse. All the stories came to life in our group collective unconsciousness as we absorbed and reacted to them.

The task of the visiting analysts was to try to fill the gaps, to bear the repetition of earlier traumas and sometimes for the therapy "to be continually aborted" as one trainee described it. The expression of what had never been and the resultant grief and rage, and of discovered love, would begin the journey to emotional health. In the trainee's countertransference, there was often a sense of being "cancelled out" and being aborted but also a wish to comfort, to hold, to weep with a weeping client. These authentic experiences helped to fill the gaps, and, as supervisors, it was pleasing to see the effects of the work and the developments in the trainees. The various compromises and adjustments we had learned to make bore fruit in intricate ways, notwithstanding the inevitable conflicts around competition and cooperation, which would follow.

South Africa

In a paper presented in Cape Town in 2007, I explored issues around South Africa's cultural complex.[32] I focused on a recurring dream of a friend and professor. The dream involved a little black boy who was playing with cowrie shells which were pretend cows. He was moving them back and forth, in and out of a pretend kraal (a round, walled container to protect cattle from predators). He was watching the little black boy. Later he realized that he wished he could be this boy and wished he could be the son of his black nursemaid, Rosie, whom he had come to realize did not "belong" in his white-skinned society because of her black skin.

In South Africa during the Apartheid years, for the ruling whites, it was quite simple—black and white literally and figuratively. The word "apartheid" itself means separateness. Boundaries were clear. Originally, this had arisen out of white economic need. In the early nineteenth century, the otherness of black people had become more pronounced when the whites, mainly Dutch, appropriated the land for cattle used by the Xhosa tribe. A series of "Xhosa Wars" followed. By this stage though, the so-called coloured race was also growing rapidly as the Dutch settlers had babies with black maids. Black skin became less strange.

But the us/them dynamics of the Apartheid ideology was entrenched. The particular boundaries set up by this ideology and the methods used to keep it in place became some of the most rigid and notorious in history. As we know, the colour of the skin indicated the positioning of in-ness or out-ness in society. As such, it played into South Africa's cultural complex in which all manner of projections were aimed at those with non-white skin by the ruling white population. Skin colour could trigger emotional reactions and was key to the cultural complex.

The dreamer had a difficult experience with his mother. He described her as "terrible." In the world of myth we know about the opposite poles of the archetypal mother. For an infant, the mother is not an individual in her own right but the "living embodiment of the Great Mother archetype."[33] She can be both creative and loving on the one hand and destructive and hateful on the other: literally, the Good Great Mother or the Terrible Mother. For our dreamer, his mother was "terrible." Remaining open to her would have been damaging for him. As a result, he would have had to set up defensive structures, or a second "skin function," to protect his undeveloped, vulnerable self from impinging experiences. Meanwhile, he had sought appropriate responses elsewhere and found them in his black nursemaid, Rosie. Rosie became the good object who stood in for mother. He said she "loved and properly mothered" him and was his "only source of unconditional love." Within these dynamics are the seeds for both idealization and denigration. The dreamer had been placed in a double-bind. In order to keep his identity within the family and to "belong" to his particular white culture, he was required to deny his experience with Rosie. He had to split off his feelings of attachment to her. These psychic splits overlaid the earlier defensive structure set up as a result of his negative experience with his mother. Complexes developed resulting from repression and splitting and consequent shadow formation. These evolved first out of his personal experience with his mother and then out of the particular culture into which he was born. His personal complexes were intricately linked to the cultural complex.

The dreamer's "required" denial of Rosie led to a positive acceptance by mother and intimate others, as well as a sense of belonging and group spirit in the various white South African reference groups in which he was involved. This sense of belonging and "in-ness" is necessary for healthy development. The negative outcome was that this aspect of denial became absorbed in the dreamer as an aspect of his shadow. There, it created a particular kind of rigidity which, in turn, linked to the rigidity of the cultural complex. The significance in the dream of the cowrie shells was about the rigidity of the shells as well as their hidden soft, feminine underbellies. The colours of the cowrie shells are the same as the hues of different human skin colour; white, caramel and black. The shells, pretend cows, had similar skin colours, but looser and softer. Cows are also generally docile and provide milk and meat. These were the creative symbols produced by his unconscious on his personal journey. They served to loosen the rigidity of his complexes so deeply linked to the culture in which he was born.

United Kingdom

As analysts in the UK, we may often remain unaware of the ways in which our cultural heritage is played out via cultural complexes. While all societies will have natural groupings around privilege, poverty, education, foreigners, etc., Britain has a particularly rigid class structure and this is a significant aspect of the British cultural complex.

Ryan researched how issues of class were played out in psychotherapy. She was concerned about the absence of frameworks for thinking and working with class and she interviewed a mix of experienced psychoanalytical psychotherapists. "Class is in you," she was told. Her exploration "lays out what is often kept politely or defensively hidden, the attributions of inferiority and superiority, which are part of class. It illustrates aspects of the construction of classed psyches." She also showed how difficult it is for therapists to use the transference and countertransference dynamics to work through such issues.[34] Schaverien similarly addressed these class issues in her in-depth work on the effects of boarding schools in the therapeutic relationship.[35]

In his recent book exploring the origins of our "classed psyches," Duffell lays bare the makings of a British cultural complex (though he did not use this term). He writes, "The political and personal can be heavily intertwined, and things that are deeply shocking stay in the collective unconscious to form the attitudes of generations."[36] He takes us back to the French Revolution, which instilled in the British ruling elite a fear of feudal uprisings. Britain's growing imperialism, together with the industrial revolution lead to what Duffell calls a capacity for objectification of others. Such objectification was clearly required to keep down the masses, colonization, and the slave trade, but Duffell's prime research has been on the effects of boarding school. The birth of the public schools (in the UK, these are high-fee private schools), he thought, was a brilliant piece of social engineering born out of a need for the "right kind of gentlemen" to run the expanding empire. He focused too on the resulting sense of entitlement. Ex-boarders, he wrote, sometimes reported that they felt like objects put through a kind of "sausage machine." Others, we know, had different experiences and for some it was a relief to have time out from problematic families; however, the public school tradition continues today to divide our class-bound society and is a lynchpin to a cultural complex.

As analysts, we naturally, perhaps unconsciously, position our patients somewhere in the class hierarchy, as they too will position us. Most of us will have attempted to join with our patients as they bring class-related themes to therapy. These might be about the struggles to free themselves from working class tentacles, humiliated by middle class smugness, or about struggles to move into the professional class. Some might feel guilty about their class privilege or envy wealthier others at their public school. Themes are played out in various ways: anxieties about accents, "good" taste, which utensils to use at meals in middle class company, etc. On the other hand, a middle class child in a free state-funded school might feel downgraded and ostracized because of his/her "posh" accent or attitudes. To varying degrees, these issues can be worked through within the therapeutic relationship. One of my patients frequently abused me because I could not possibly understand what she had been through. "You have obviously had such a privileged background," she insisted. Exploration of patients' class perceptions of the analyst and vice versa needs attention in therapy and the analysts' awareness of their own prejudices and class will help loosen the conflicts.

Aside from class, there are other groupings of immigrants and the degree of prejudice and "objectifying" is continually in flux.[37] Our immigrant patients, too, will experience issues in relation to the in/out dynamic and cultural complex. A Jewish patient once came to see me assuming I was Jewish. Her shock to discover that I was not was very useful in the therapy. Freud himself was an immigrant, grateful for the generosity of the British who opened their boundaries to Jews fleeing the Nazis. It is difficult not to be aware today of the ambivalence, in some circles, about admitting the "foreigner" to the UK. Because I am an analyst who is a "foreigner/outsider," my patients can be released from anxieties about class, but other reactions can be stirred emanating from a cultural complex. Many years ago, arriving in England as a white South African at a time when Britain was boycotting South Africa, it was often assumed that I was racist. Meanwhile, when back in South Africa, I had sometimes been accused of being a communist and "native lover." Different countries at different times reveal different cultural complexes.

Concluding Remarks

The focus of this paper has been on the dialectical relationship between the culture and the individual and how different histories play into both personal and cultural complexes. A greater understanding of these processes is vital now that there are so many new cross-cultural projects alive in countries and cities with different cultural legacies to our own.

It is worth holding in mind our own missionary and colonial backgrounds and the "righteousness" of some "missions." These are inevitable aspects of the cultural complex of the West. Missionaries took Christianity to far-flung areas of the world with a hegemony that now makes us shudder. Similarly, in our colonial and imperial past, we were able to "conquer" and "objectify" the native, disregarding their traditions and needs. Our project now is a very different one because we have been invited by other countries to teach, but we still need to be conscious of our roles of "privilege," and "entitlement" and the dangers of supremacy hidden within our own cultural complexes.

Each school of psychotherapy or analysis has its own culture, its own sense of righteousness and its own complexes. Keeping this in mind is helpful when confronted with challenge and resistance to "our way" of working when abroad. As analysts, we are required to imagine ourselves into the world of our patients. We do well to imagine ourselves into the cultures where we are guests. In this way, we are more able to bond and find creative solutions to free up those from the host culture to find their own ways.

Notes

1. Zygmunt Bauman, *Liquid Times* (London: Policy Press, 2007).

2. Farhad Dalal, "Jung, a Racist," *British Journal of Psychotherapy* 4 (3, 1988): 263–279; and Andrew Samuels, "National Psychology, National Socialism, and Analytical Psychology: Reflections on Jung and Anti-Semitism," *Journal of Analytical Psychology* 37 (1, 1992): 3–28.

3. Helton Godwin Baynes, *Mythology of the Soul* (London: Bailliere, Tindal and Cox, 1940).

4. Joseph Henderson, *Shadow and Self: Selected Papers in Analytical Psychology* (Wilmette, IL: Chiron, 1990), p. 104.

5. Michael Adams, *The Multicultural Imagination: "Race," Color, and the Unconscious* (London and New York: Routledge, 1996), p. 40.

6. Orlando Figes, *Natasha's Dance: A Cultural History of Russia* (London: Penguin, 2002), p. xxxiii.

7. Richard Dawkins, *The Selfish Gene* (Oxford: Oxford University Press, 1976).

8. Iain McGilchrist, *The Master and His Emissary: The Divided Brain and the Making of the Western World* (London and New Haven: Yale, 2009), p. 247.

9. Jolande Jacobi, *The Psychology of C. G. Jung* (London: Routledge, 1942).

10. Zygmunt Bauman, *Liquid Times*.

11. Thomas Singer and Samuel Kimbles, *The Cultural Complex: Contemporary Jungian Perspectives on Psyche and Society* (Hove and New York: Brunner-Routledge, 2004).

12. Henderson, *Shadow and Self*, pp. 103–23.

13. C. G. Jung, "The Structure and Dynamics of the Psyche," (1931) in *The Collected Works of C. G. Jung*, vol. 8, trans. R. F. C. Hull (London: Routledge & Kegan Paul, 1960), § 203.

14. Singer and Kimbles, *The Cultural Complex*, p. 4.

15. Thomas Singer and Samuel Kimbles, "The Emerging Theory of Cultural Complexes," in J. Cambray and L. Carter, *Analytical Psychology* (Hove and New York: Brunner-Routledge, 2004), pp. 176–203.

16. Joseph Redfearn, "Movements of the I in Relation to the Body Image," *Journal of Analytical Psychology* 39 (3, 1994): 283–310.

17. C. G. Jung "The Psychology of Transference" (1946), in *The Collected Works of C. G. Jung*, vol. 16, trans. R. F. C. Hull (London: Routledge and Kegan Paul, 1954).

18. Anthony Stevens, *Archetype: A Natural History of the Self* (London: Routledge and Kegan Paul, 1982), pp. 210–43.

19. *Ibid.*, p. 211.

20. Bruce Chatwin, *The Songlines* (London: Vintage, 1988), p. 201.

21. Rosemary Gordon-Montagnon, "Do Be My Enemy for Friendship's Sake," *Journal of Analytical Psychology* 50 (1, 2005): 27–34.

22. Michael Fordham, *Explorations into the Self*, vol. 7, *The Library of Analytical Psychology*, eds. M. Fordham, R. Gordon, J. Hubback, and K. Lambert (London: Academic Press, 1985).

23. Freud (1927), as cited in Brian Feldman, "A Skin for the Imaginal," *Journal of Analytical Psychology* 49 (3, 2004): 288.

24. Esther Bick, "The Experience of the Skin in Early Object Relations," *International Journal of Psychoanalysis* 49 (1968): 484–86.

25. Didier Anzieu, *Psychic Envelopes* (London: Karnac Books, 1990), p. 19.

26. Brian Feldman, "A Skin for the Imaginal," *Journal of Analytical Psychology* 49 (3, 2004): 285–312.

27. McGilchrist, *The Master and His Emissary*, p. 250.

28. Alexander Etkind, "How Psychoanalysis Was Received in Russia," *Journal of Analytical Psychology* 39 (2, 1994): 191–202.

29. Catherine Kaplinsky, "Soul on Ice, Soul on Fire: Abandonment, Impingement, and the Space Between," *Journal of Analytical Psychology* 37 (3, 1992): 299–321.

30. Rushi Ledermann, "The Infantile Roots of Narcissistic Personality Disorder," *Journal of Analytical Psychology* 24 (2, 1979): 107–26.

31. Elena Bortuleva (personal communication, 2014).

32. Catherine Kaplinsky, "Shifting Shadows: Shaping Dynamics in the Cultural Unconscious," *Journal of Analytical Psychology* 53 (2, 2008): 189–207.

33. Stevens, *Archetype*, p. 89.

34. Joanna Ryan, "Class Is in You: An Exploration of Some Social Class Issues in Psychotherapeutic Work," *British Journal of Psychotherapy* 23 (1, 2006): 49–62.

35. Joy Schaverien, "Boarding School: The Trauma of the 'Privileged' Child," *Journal of Analytical Psychology* 49 (5, 2004): 683–705.

36. Nick Duffell, *Wounded Leaders: British Elitism and the Entitlement Illusion* (London: Lone Arrow Press, 2010), p. 160.

37. Salman Akhtar, "The Mental Pain of Minorities," *British Journal of Psychotherapy* 30 (2, 2014): 136–53.

CHAPTER 6

UNDERSTANDING GROUP AND ORGANISATIONAL DYNAMICS IN CULTURAL PARTNERSHIPS

STEPHAN ALDER

INTRODUCTION

In this chapter, I would like to describe and discuss some personal, group, and collective dynamics emerging from the work I have done with foreign colleagues in other countries. As a psychodynamic psychotherapist trained in group analysis as well as a Jungian psychoanalyst and member of the International Association for Analytical Psychology (IAAP) from Germany, I have worked with colleagues both individually and in groups and organisations over the past ten years in Russia, Kazakhstan, and Switzerland. Based on my experiences during this time, I hope to address the following questions:

- What can we learn about cultural partnerships from a study of the theory and practice of group analysis working with small groups and in organisations?
- Who am I and who are "the others?"
- What kind of projections do I experience and receive from "the others?"
- What do I project onto "the others?"
- How do they respond to me?

- What are the opportunities and vehicles in which we can reflect together on these processes?
- How to create a space for my own reflexive function?

Firstly, I would like to give a brief introduction of some basic group analytical theories and organizational dynamics and then I will describe my impressions as a German (actually an East German) analyst working in Russia and Kazakhstan, both part of the Soviet Union from 1922 until 1991. I will also compare these impressions with my experiences working in Switzerland. Secondly, I want to explore the fear and shame I felt in front of groups of colleagues abroad. As psychoanalysts and group analysts, we know that at a first meeting, levels of anxiety are usually high, combined with a certain curiosity about new experiences, and new faces in different cultures. When Germans go to Russia, however, there is often a higher level of collective fear and shame (conscious and unconscious) because of what happened historically between Germany and the Soviet Union in wartime. Thirdly, I would like to give some examples from the clinical seminars and group work where there were differences in gender roles. Finally, I will try to explore some questions and tasks for the future. These can, of course, be seen as nothing more than personal descriptions; however, they may stimulate further research, or encourage colleagues to do similar work. Earl Hopper, in his thinking about organizations, makes the following points:

> Usually an "organization" refers to a complex social system; that is, one with a high degree of role differentiation and specialization; with a large(ish) number of members that has been chartered or at least recognised by other people; often planned and designed intentionally by the founders of it in order to achieve certain goals more effectively and more efficiently than can be achieved by a small(ish) number of people who are merely associated or loosely connected with one another, and who lack a *modus operandi*.[1]

Interest in organisations and group functioning and the search to understand their inner dynamics seems to be growing. The opportunity to carry out a comparative qualitative investigation of groups and how they organize themselves in two or three different countries is a project that promises to be particularly interesting in terms of group dynamics. The quotation from Earl Hopper gives a detailed description of what an organization actually is. Hopper apologized for this long definition, but he also expressed within it the complexity that goes hand in hand with thinking about organizations and the dynamics within them.

As group analysts and Jungian psychoanalysts, we have a conviction about the impact and power of a dynamic unconscious. This dynamic unconscious unfolds both within individuals and between individuals living in connection and communication with partnerships, groups, large groups, organisations, and cultures. Communication is connected with mutual projections and identifications between individuals and groups. It can be seen in every individual. The vertical dimension is the notion of depth from the personal conscious and unconscious to the deeper levels of large groups to the level of the collective unconscious. During a seminar he gave in 1925, Jung painted a "geology of personality." The so called geological levels are the following: "A = Individuals; B = Families; C = Clans; D = Nations; E = Large Groups (European man, for example); F = Primate Ancestors; G = Animals (Ancestors in general); [and] H = "central Fire."[2] We can understand these levels described along a vertical dimension as the mirror of the horizontal reality and impact. The horizontal dimension is the network between individuals which creates groups like families, large groups like clans, organisations, and cultures. The role of time and history means that we can appreciate our ancestors, whom we (re)find in families, large groups, organisations, and cultures. A hierarchy of differentiation may be observed as complex because of the interaction between the vertical and horizontal dimensions, each offering different hierarchies. If we try to imagine this multiplicity, we can visualise an invisible network or matrix.

The concept of a psychodynamic matrix as a complex communicational network was first described in 1944 by one of the most important pioneers of group analysis, Siegmund Heinrich Foulkes.[3]

> The matrix is the supposed communication and relational texture in a given group. It is the shared common background that ultimately determines the meaning and importance of all events and on which all communications and interpretations, both verbal and non-verbal, are based.[4]

Foulkes differentiates between the levels of the current personal conscious, the unconscious transference dynamics, and the primordial/archetypal. This concept of a matrix as a communicational network is a model which is relevant intra-psychically as well as inter-personally. It is also used to describe a state of being and a transitional process in time. The unconscious resonates in individuals and in groups along both a horizontal and a vertical axis. The unconscious may be like a resonating organ.[5] When we focus on organisations, we can use the matrix metaphorically to capture the invisible network of relationships and mutual projections. This creates a power that either holds us together or disconnects us, splitting us into groups or subgroups.[6]

I came to work as a group analyst abroad through my activities in the IAAP and its German Society, the German Society for Analytical Psychology (DGAP). It began in 2003 when my Swiss colleagues asked me to run a Balint group in Zürich. This developed into a programme of on-going cooperation in the form of group supervision on two Saturdays a year until 2014. Balint group work was developed and disseminated by Michael Balint (1896–1970) in the 1950s and 1960s. He was a Hungarian psychoanalyst living in London. In his case seminars he focused on the doctor-patient relationship.[7] The work was done in small groups of between three and twelve people where one colleague would be invited to bring a case. Everything that happened in the group was viewed as connected and related to the case, especially the doctor-patient relationship.[8] In the group, an intersubjective room could be created through the group's communications. With the help of a clear frame (in terms of time, location, confidentiality, etc.) and techniques of free association, active imagination, reflection on scenes, and amplification, the complexity of the transference/countertransference dynamics between the group members could be understood as a mirror of the patient/therapist relationship.[9] In this way group members and the person presenting the case were able to gain more conscious understanding of the dynamics of the case.

These have been the models of work I have used as a visiting group analyst in Switzerland, Kazakhstan, and Russia where I have given theoretical seminars and clinical supervision in small groups. This is the ongoing work on which I want to focus in this chapter which will, I hope, bring to life my personal approach to working in different cultures.

Working in Kazakhstan

Kazakhstan was founded in its present form when the Soviet Union fell apart in 1991. In terms of land mass Kazakhstan is the ninth largest country in the world but has a mere seventeen million inhabitants. A small part of the country is located in Eastern Europe, while the larger part, including the old capital city of Almaty (Russian: *Alma-Ata*) and the newly-founded capital Astana, are part of Asia. Around 1.4 million people live in Almaty and the population is made up of many different nationalities—Kazakhs, Russians, Uzbeks, Uighurs, Tatars, Germans, and many more. It is already many years—or it feels like many years—since I was asked by an older colleague if I could imagine flying to Kazakhstan to give a theory-based seminar in analytical psychology. There was, he said, a group of colleagues in the Kazakh city of Almaty who wished to train as Jungian analysts and who were involved in a programme of study associated with the IAAP Router Programme. I had already heard of Kazakhstan and Almaty a long time ago at school and I decided almost immediately that I would go. After a period of about a year everything was in place. A Kazakh colleague contacted me via email and told me that she had handed over the management to someone else. This new colleague contacted me soon afterwards and we agreed on the details for my first visit. This significance of this change of who was in charge and the potential conflict it would generate in the group was something I only began to understand later. When I arrived in Almaty, spring had just arrived, the snow was gone and the sun was already bringing warmth to the air. I could see a long mountain range, the Altai Mountains, from the plane. After landing I was welcomed by two friendly colleagues. I had been quite anxious about the trip—worried about being over-controlled, perhaps even being interrogated and prevented from entering the country. The reason for these threatening internal scenes in my imagination was based on past experiences with unpleasant border crossings when travelling from the German Democratic Republic (GDR, 1949–1990) into other countries such as Poland, the former Czechoslovakia, the Soviet Union, or the former Federal Republic of Germany. Fortunately, none of the terrible things I was imagining occurred and I passed safely through border control at the airport with a stamp in my passport and a special residency permit. My reception was friendly and very welcoming. My

Kazakh colleagues were used to receiving foreign lecturers and they showed me to a three-roomed apartment where I was to spend the following four days. I met the group on the first seminar day. There were about ten women and one man; a second man joined the group later. The group's core members were women who had not only rented rooms in an office building shortly before I arrived, but had also refurbished them. The three days of seminars, each with four ninety-minute teaching units, was a format similar to one I was used to in Germany. A likeable interpreter translated for me and she managed the subtleties of language expertly in both English and Russian.

At first, I established the organizational setting for our partnership. This meant clarifying the place, times, how long the sessions would be, the breaks, obligatory presence of the participants, cancellations, payment issues, and issues concerning confidentiality about personal information, patients, and institutions. For this, I ask participants for their consent and also if necessary, to make some changes. The group evolves within this given setting.[10] Further, there is an unconscious process of development. Tuckman summarizes the stages of team development as the following: forming (orientation phase); storming (confrontation phase); norming (cohesion phase); performing (growth phase); and adjourning (termination phase).[11] This is very similar to MacKenzie whom I will refer to later in my text.[12] In the following account, I will describe in my own words how I make use of these steps during my work.

First of all, clarification is made about who belongs to the group, who doesn't belong, who is still to come, who won't be coming, who is new to the group, and who belongs to the core group. This creates an important boundary in terms of the setting and negotiates the inner-outer boundary of the group that is forming. This is the first important input to the seminar units. In the following sessions, issues concerning power relationships within the group arise during the work process (storming). Who is in charge of the group, who is subordinate, who is a rebel, who is allowed to turn up late, and who prefers not to say anything? Once this process has settled (norming), the third level to negotiate is about intimacy (performing, closeness-distance). How open can I be, how personal can I be when contributing to the group sessions, and how can I establish my own limits including what I can keep to myself? At the end of the three-day seminar, I was told by the members of the group that they felt they were able to experience themselves as a group in a particular way (more cohesive fantasies emerged). Enabling and improving the communication in a group is one of the main tasks for every group leader.[13] This obviously evolves as a joint process in the group and strengthens the feeling of togetherness.

I was told that the group had recently gone through a crisis. The background to this was that one colleague had already completed her training to become a Jungian analyst. This had put her in the role of the pioneer. She had developed something important for herself and she had also helped to motivate other colleagues to take up analytical training; however, she was not interested in leading the study group that was evolving. Why should she, one could ask? It is not an obligation, but some of the new members expected it and this led to a feeling of annoyance and disappointment with her. Because of her decision, those coming after her were forced to reorganize themselves, which they did. I was happy to see how well they managed to evolve a new shape to the group as well as the process of working together to manage the crisis.

This is one model for organizational development. Initially, one or two people begin something new followed by others and the emergence of a group culture. The first members might become the founding figures of a group or they might merely provide an impulse for group activity and then go their own ways. The new members of a group then develop a culture of togetherness, thus creating a new and more complex dynamic structure.

I travelled to Kazakhstan a second time, one year later. The study group was active and various lecturers from different countries had come to give seminars. This time, the group work was intense in another way—you could say it had evolved. I organized the daily structure so that I taught theory in the first three units and offered a supervision session in the fourth unit. Here, a patient or a group could be presented. One of the students presented her patient group in the psychiatric clinic where she worked. Severely ill patients there were willing to speak with one another in a small group and this student proved to be very competent. On the following day, another student in the group—a psychologist—presented his outpatient group, and it was also good. On the third day, these two presented the group they ran together at a day clinic. During this presentation, only the male colleague spoke and it was as if the female colleague had ceased to exist. When I pointed this out, I was told that this reflected their understanding of male and female roles. When a woman is alone, she is strong, they said, and so is a man; however, when a woman is together with a man, then she subordinates herself to the man and becomes invisible. This was an interesting observation which revealed itself both consciously and unconsciously, but to me it jarred with the idea of equal roles for men and women. The female Kazakh colleagues also realized this, as did the two men actually. The participants in the seminar reflected together on this phenomenon, seeing it as a point of tension between something traditional and more modern. On the one hand, they found such behavioural roles normal, but also agreed that it didn't necessarily have to be that way. And then, to ease the situation, they laughed about it. To what extent these gender-specific role images influence the inner representation in a group would be an interesting issue to research further.

Visiting Krasnodar in Southern Russia

I needed to check on the map to find where Krasnodar was located. I was amazed to find out that it was very close to the Black Sea, close to the Ukrainian border, and that it is a large and prosperous city. It has a population of 745,000 people, and Pushkin and Tchekhov had been there. The city was founded by Catherine the Great in 1793 as a bastion against the Ottoman Empire. The city was named Ekaterinodar first and then renamed Krasnodar in 1920, the "red city" after the 1917 Revolution. For some decades, the large statue of a man, Vladimir Ilyich Ulyanov (alias Lenin), was to be seen. At the break-up of the Soviet Union, the Lenin memorial got a counterpart. This was the statue of a woman: Catherine the Great. A great woman and a great man could see each other. After probably three or four years, the Lenin statue was removed and its base is now empty. Sometimes a fountain is there. Catherine the Great, however, continues to welcome visitors to Krasnodar, and young couples often have themselves photographed with her on their wedding day.

In cooperation with a colleague from Moscow and another colleague from Kiev, we offered group supervision sessions for trainees involved in the IAAP Router Programme. There were twenty-one participants in total and we worked in smaller and in larger groups. In this way, everyone was able to present something from his or her ongoing clinical practice that was part of their training. I would like to present an example:

My supervision group was made up of ten female colleagues and two men as well as an interpreter. It was the third session. One of the participants presented a female patient aged thirty, who had been sexually abused by an older man when she was twenty and then went on to marry him. She lived with him for six years. The patient came for therapy because although she was now divorced, she was depressed, felt out of touch with herself, was sometimes aggressive, and shouted a lot. She said that she found it hard to find a place for herself in the world. The patient came for therapy twice a week. Her mother refused to take her seriously, especially about anything to do with feelings. The patient described a dominant and demanding mother and a weak father. The candidate wanted to understand better her relationship with her patient. She noticed that she was only able to feel a connection with her patient when they were saying goodbye at the end of a session.

The group experience led to the following associations and resonances:[14]

- hypnotic reactions and sleepy states
- the picture of a hole—the hole became larger and became a frightening vortex
- a feeling of tangible danger as well as the impulse to flee, "to get out of here!"
- the perception of the efforts made by the two women (patient and therapist) to make closer contact with one another despite difficulties
- questions of who was abusing whom

There were also descriptions of physical coldness, anger, emptiness, an emotional void, and frightened eyes of group members. The emptiness and the atrocity were received and perceived in the group as elements of "nameless dread."[15] Later, we could name these bodily reactions and feelings as scenes of severe trauma and we could explore them psychologically. The group was able to work as a container for a possible transformative process. Jung conceptualized such a transitional process as the transcendent function and Bion conceptualized it as alpha function.[16] At the end of the session, the woman presenting the case told the group the following biographical information: the patient's grandmother had been raped by a German soldier in 1942 in southern Russia. In order to survive, she stayed with this German soldier in Krasnodar during the seven-month period of German occupation (from August 1941 to February 1942). There was no information about what became of her or whether she survived the German withdrawal in 1942, however, everyone experienced this as a shock but also as revealing. This same pattern was recognizable in the patient: first sexual abuse, then continuing to live together. It seemed as if this kind of subjugation had been passed down the generations as a traumatic pattern. It was a message to me. Was I the aggressor here, the rapist, who simply satisfied his primitive needs without shame or guilt? How could I position myself in relation to my collective, transgenerational past? I tried to respond, and spoke of my humiliation, my feelings of guilt, and my disgust. Taking on this large group identity as a German was an important step both for me and for the group in terms of their attitudes toward me.[17] Could I acknowledge—even at the end of the session—the horror, the pain, the continuing injury inflicted by my grandparents' generation? I had tried to express this in words. In my mind I saw frightened looks unable to grasp the horror of it and tears came to my eyes. This became visible to the group and could be put into words. The words were translated from English into Russian, then from Russian into English, and, in my mind, partially translated into German. The English word *atrocity*, referring to "nameless dread" and terrified horror, had already taken on its own meaning.[18] In Russian the words are *ujschas* and *sverstwo*.

The next day I was shown around the city. During the tour the chief psychiatrist told me that in 1941 the Germans first of all shot the mentally ill, then the Jews. It was only after that that the Russians were deported to German work camps to do forced labour. I felt cold all of a sudden. I felt angry, ashamed, and sad. After painfully acknowledging my German origins, the psychiatrist was able to mention some of the terrible things that happened during the Stalinist era. This was the beginning of a powerful psycho-historical exchange between us.

Working in Zürich, Switzerland

Switzerland is a small country in the middle of Europe. It has eight million inhabitants, 366,000 of whom live in Zürich. The landscape is dominated by the Alps and clear, beautiful lakes. As someone who spent thirty years of his life in the confinement of the GDR, I was unable to travel to countries in the West. A school friend of mine had fled the GDR with her parents in 1969 to go to West Germany, living for a period in Zürich. It was always a place I had wanted to visit. The fact that Jung lived and worked there for a long time was also significant, of course. The border towards the West opened in 1990 and I enjoyed travelling to Switzerland finding well-kept towns and villages which were generally better tended than in Germany.

My analytic colleagues there received me in a friendly and welcoming manner. For them, I came from Germany, from Berlin, and it wasn't until later that they asked me if I also knew East Germany. I offered supervision and this seemed to work well. The questions about whether I came from East or West Germany, and what that actually meant, did not seem to be as relevant there as I experienced it during my work in Germany.

I would like to now make some theoretical comments about group work. When supervision begins, every group follows unconscious rules and tasks. MacKenzie delineates four stages of a group process.[19] This model may be compared with Tuckman's, as referred to earlier. MacKenzie's first stage is named "engagement" (similar to forming). Here the group looks for an understanding of how to work together. It is like a warming-up. All participants wish to find a shared path or possibly the opposite. The main task for the group leader is to support an atmosphere of cohesion. The second phase is "differentiation" (similar to norming). This stage observes how group members want either to show how good they are or someone wants to blame another group member. Therefore, the task of the group leader or group conductor is to integrate the negative and disruptive tendencies into the group functioning and to relate this to the therapist-patient relationship. The third stage is "interpersonal work" (similar to performing). The group is able to work and to communicate in a creative way. Members are able to open up towards each other to a greater degree and more intimacy is possible. The last stage is "ending" (similar to adjourning).

During a one-day course it is always clear that the group has to deal with feelings of sadness, disappointment, and mourning. The fantasy of meeting again can be shared.

MacKenzie quotes Schutz's three stages in the group in his own work and has used these to develop his own ideas.[20] Schutz's stages move from inclusion, to control, and then to intimacy. I will refer to them later in the chapter but I am drawn to the following guidelines in my own work as they constitute a more flexible model:

> Stage 1. forming/engagement/inclusion—using engagement to form and include
> Stage 2. norming/differentiation/control—using the process of differentiation to understand the norms of the group
> Stage 3. performing/interpersonal work/intimacy—using performing which facilitates intimacy by supporting the interpersonal work
> Stage 4. ending/adjourning—using the remaining time for concluding remarks, experiences of loss and sadness, to say goodbye

I would now like to present the group work done with my colleagues in Switzerland. I have worked with this group in Zürich as a supervisor and Balint group leader for seven years. We meet up twice a year and the group members have given me permission to present and discuss the following case material. The group had nine members on the day of the session, eight women and one man; two others were absent. We were in the middle phase of the supervision process, concerned with acceptance and rejection and/or complications with issues of unconscious control and intimacy. The following is the case example which the group has kindly given me its permission to present.

A German man had fallen in love with a woman living in Zürich. She was hesitant and unsure if she really loved him. I took this theme as a group unconscious fantasy in relation to me as their group analyst. The psychotherapist was able to gain access to her previously unconscious negative maternal countertransference. What could this have to do with group and organisational dynamics? My thoughts were that the supervision group carried this ambivalence within itself and worked casuistically via the choice of case. I understood from this case presentation that the group felt hesitant and unsure whether they liked working with me as their group leader. I felt like the young man who was trying to come closer to the group. Through sharing our thoughts and feelings, the group members and I were discovering more affection and intimacy towards each other.

As a second more detailed example, I would like to share another group session with the same group. One of the group members presented a twenty-five-year-old female patient with severe depression and an unhappy love life. The patient's mother called the therapist on the phone to ask for psychotherapy for her daughter. The daughter was depressed, she said, and very unhappy in her life. The patient came to an appointment. She was very tall, thin as a rake, and with long, blonde hair. She smiled a lot. The female therapist was having difficulty making any kind of contact with her patient. The patient had met a young man from another city about three hours away. He had fallen in love with her, though she was more reluctant and did not fall in love straight away. It was as if he became fixated on her. She refused his love and yet was pleased when he came to see her. She was at university but found it difficult to concentrate because of her depression. The therapist and patient have had twelve sessions to date. The patient's problem was that she could not find a way to let her boyfriend know where her boundaries were. She told the therapist that he would call her on the phone and then a few hours later, she would find him standing outside her window in Zürich. She said she had lost six kilograms in weight, but had regained two kilograms since beginning therapy. The group member who was presenting thought this was a good sign. There were some questions from the group, after which there was an agreed period of silence for contemplation. When this silence was over, the group members began to talk slowly.

> Ms. 1: I wasn't able to concentrate and I had to sneeze at the beginning. The sneeze felt like a release. I soon began thinking about the patient and her inability to concentrate, but I still didn't understand anything that was going on.
> Ms. 2: An image came to me. At first it was vague, then I saw a figure in a cave like a whale. I thought about Pinocchio and how his nose grows and stabs the

whale. That's how he got out—also a kind of release. (She laughed.)

Ms. 3: My chest feels tight. It's really tight and very unpleasant. When I heard about this young woman, I felt I like her. It's the heart. (She falls silent.)

Ms. 4: I can't feel anything. I have no feelings at all. There is a mask there. I see a smiling face. I think about the smile and feel nothing. As I am saying this, I can imagine that this might be how the patient feels. It feels strange and disconcerting.

Ms. 5: She's like a fairy, so delicate, transparent. The young man is standing there, always turning up again and again. He provokes her, looks for her but he doesn't reach her, although he sees her in the way you see fairies.

Ms. 6: I saw a spider's web before my eyes, a well-spun web. It was shimmering in the wind. The patient is tangled in the web and yet still free. Her parents are pulling at the strands of the web and holding on to them. And the therapist is standing beneath it saying, "Let yourself fall, I'll catch you and hold you." Maybe it hasn't reached her yet, but she can see it. (Comment from me: In Ms. 6's statement we can see in a wonderful way, the network of the unconsciously active parts of the surrounding group and the other important people.)

Ms. 7: It's wooden and stiff. It feels like a wooden board. Then I thought about the mother-daughter relationship. Perhaps the twenty-five-year-old girl has difficulties saying *no*. But she can say *no* to the man. He asked her to marry him and she said *no*. That's also an achievement, isn't it?

Ms. 8: Yes, I had that idea as well. She is so thin but there is defiance there. She is able to say *no*.

Ms. 2: I talked about Pinocchio. I was thinking about being in love. He wanted to fall in love. He was still wooden and yet he felt something, maybe in the way a fourteen-year-old loves.

>Mr. 1: I've been thinking the whole time about the young man. He risked so much, and came from some distance away to find her. He told her he loved her, asked her to marry him, and tried to surprise her by suddenly appearing below her window. He has so much energy. And she, the depressed patient, triggered it off. That's important, isn't it? I'm impressed somehow, and yet the relationship still isn't working. That makes me feel sad.

During supervision groups or Balint groups, it regularly happens that different aspects of the patient, of the case presenter, and their relationship to each other are transferred into the group and expressed consciously or unconsciously by group members. In this case, the group experienced—in a similar way to the patient, the analyst, and their relationship—the feeling of being tangled up in a web; we felt the same kind of paralysis, the woodenness, the mask-like feeling, the longing for relationship, and the lack of an emotional response. It is often a surprise to the presenter how accurately group members perceive what is going on in the therapeutic relationship. Even that which remains unsaid sometimes finds expression through ideas, images, or physical reactions from the group members. We can understand these group phenomena through the concept of empathy, with the help of the neural correlate of mirror neurons.[21] Previous models of resonance and polarization, such as those described by the psychoanalyst and group analyst S. H. Foulkes, convey something similar.[22] With her theory of complexes, Verena Kast described the process of apportioning parts of complexes and complex episodes to the analyst and the patient.[23] Jung said that in order to really be aware of the patient as a system himself, one has to be aware of the impression of the soul in one's own soul, in one's own system as a "reciprocal interaction between two psychic systems."[24] In other words, we can speak of a complex-like apportioning of parts of a system from the patient into the field of the supervision group. Everything that happens and is said in the group has to do with the matter in hand and with the therapist-patient relationship. The group as an intersubjective space comes into being through the unconscious and conscious interactions of its group members. This is where feelings, thoughts, images, and spontaneous fantasies or physical reactions take place. The group, the group process, the relationship to the group leader, the infringements of the rules, and the organizational aspects are consistently understood and interpreted in relation to the case. They can also be understood and used in other ways, of course, but this can then overstep what has been agreed upon within the group.

Later in the group, one colleague, Ms. 3, maintained that she was unable to get a feel for the patient, and yet slowly a picture of her was emerging in her mind. She said she felt some kind of relationship to her even if she still seemed to be quite far away. Ms. 4 confirmed that she had had some thoughts about the patient's mother and father, about whom hardly anything had been said, and about what kind of family she might have. Mr. 1 thought once again about the young man and how creative his loving feelings were.

I summarized what had been said and commented that I wondered why we had heard so little about the relationship between the two women, the patient and the therapist, comparable to daughter and mother. All we had really been told was that the patient was able to let herself fall, but didn't know it yet. It was significant that the group could not get a feeling for the patient. It felt important to understand the patient and what was going on in the therapist. The therapist was unable to make contact with her patient and the patient's problem, in turn, was that she was unable to establish personal boundaries for herself. This is something that also played an important role in the here and now of the group supervision session.

At this point, I turned to the presenter and asked her to comment on the perceptions, reactions, and thoughts from the group. She was happy to have received so much feedback. Much seemed familiar to her, especially her difficulty of not being able feel with her patient. She also commented that the image of the spider's web was helpful, adding that the patient really did feel entangled, as if everyone was tugging at her and she was no longer able to move. She also liked the idea that the patient could let herself fall out of the spider's web and could be caught metaphorically and held by the therapist. She hoped this would now become easier. She was now thinking more about a little girl, which seemed strange and there was some resistance in her to this idea. We discussed the nature of this resistance. Was it resistance against the idea of a mother and a small child? Did the mother-therapist not wish to understand (metaphorically catch and hold) her little girl-patient? What kind of resistance and rejection were we seeing here? We considered and discussed it. In doing so, something changed in the group feelings about the case.

It was clear that the young woman came across as very child-like and that she did not receive enough attention and love. She was becoming aware of the idea that she may not be loved and how her parents pressured her to perform well at school. This brought about a change and a stronger feeling of relatedness as well as sadness. The pain and feeling of tightness in the chest area became less for some group members. The fairy figure acquired more substance, began to feel real, and started to seem more like a woman. The relationship between the therapist and the patient changed and in such a way that the therapist was now curious to see what would happen when the patient came the following week. Confidence grew where before there was only a feeling of searching for something. On the one hand, the presenter and the group members were preoccupied with the nature of the therapeutic relationship; on the other hand, each of us was trying to relate to the group and to what had been described. Hope of a developing relationship grew, in contrast to an earlier group experience of a desperate searching. It was something we could feel and think about together as a group.

Conclusions

The work I have done with colleagues in these different cities has shown me a diversity of different ways of living while at the same time, similarities in the human condition. Analytical psychotherapy can facilitate respect, patience, and humane responses within the therapeutic encounter. The study groups which emerged in this way went through typical developmental phases of a group.[25] Each group, of course, has its individuality as do the members who make up the group. What is special about this joint work process is that as teachers and learners we are from the very beginning part of a whole network of experiences and relationships. We are part of a group matrix, the invisible network of relationships described by Foulkes.[26] As such, bilateral and international connections are consolidated and these are meaningful for our profession and for us as human beings. It is always a joy for me to participate in such life-affirming group development processes. I would like to close with a quote from Gerhard Wilke:

> The message is, nothing gets done alone, and all outcomes depend on interdependence. In contrast, the prevailing myth of the age is: it is all up to you alone and it all depends on the leader.[27]

Notes

1. Quoted in Gerhard Wilke, *The Art of Group Analysis in Organisations: The Use of Intuitive and Experiential Knowledge* (London: Karnac Books, 2014), p. ix.

2. C. G. Jung, *Analytical Psychology: Notes of the Seminar Given in 1925*, ed. William McGuire (Princeton, NJ: Princeton University Press, 1989), § 1.

3. S. H. Foulkes, *Introduction to Group Analytic Psychotherapy: Studies in the Social Integration of Individuals and Groups* (London: Karnac Books, 1991).

4. S. H. Foulkes, *Therapeutic Group Analysis* (London: Karnac Books, 1964).

5. Günter Gödde and M. B. Buchholz, *Unbewusstes* (Gießen: Psychosozial-Verl, 2011).

6. Wilke, *The Art of Group Analysis in Organisations*, p. ix.

7. Michael Balint, *Der Arzt, sein Patient und die Krankheit* (Stuttgart: Klett, 1957).

8. *Ibid.*; Michael Balint, "Forschung in der Psychotherapie," in *Fünf Minuten pro Patient* (Frankfurt: Suhrkamp, 1975), p. 35; H. Argelander, "Balint-Gruppen—ein Fortbildungs- und Forschungskonzept," *Die Balint Gruppe in Klinik und Praxis* (1, 1988): 58; and Stephan Alder, "Durch aktive und geleitete Imagination modifizierte Balint-Gruppenarbeit," *Balint Journal* 4 (1, 2000): 114.

9. Wilke, *The Art of Group Analysis in Organisations*, p. ix.

10. I. D. Yalom, *Theorie und Praxis der Gruppenpsychotherapie: ein Lehrbuch*, vol. 66 (Stuttgart: Klett-Cotta, 2007).

11. Philippa Perry, *Couch Fiction: wie eine Psychotherapie funktioniert* (Kunstmann, 2011); and Bruce W. Tuckman and Mary Ann C. Jensen, "Stages of Small-Group Development Revisited," *Group & Organization Studies* 2 (4, 1977): 419.

12. Tuckman and Jensen, "Stages of Small-Group Development Revisited," p. 419.

13. Foulkes, *Introduction to Group Analytic Psychotherapy*.

14. *Ibid*.

15. Wilfred R. Bion, *Erfahrungen in Gruppen und andere Schriften* (Stuttgart: Klett-Cotta, 2001); and Wilfred R. Bion, *Experiences in Groups* (London: Social Science Paperbacks, 1961).

16. *Ibid*.

17. Vamik Volkan, *Blind Trust: Large Groups and Their Leaders in Times of Crisis and Terror* (Charlottesville, VA: Pitchstone Publishing, 2004).

18. Bion, *Experiences in Groups*.

19. Tuckman and Jensen, "Stages of Small-Group Development Revisited," p. 419.

20. *Ibid*.

21. Joachim Bauer, *Warum ich fühle, was du fühlst. Intuitive Kommunikation und das Geheimnis der Spiegelneurone* (Hamburg: Hoffmann und Campe, 2005).

22. Foulkes, *Introduction to Group Analytic Psychotherapy*.

23. Verena Kast, "Komplextheorie gestern und heute," *Analytische Psychologie* 29 (4, 1998): 296.

24. Jung, *Analytical Psychology*, § 1.

25. Foulkes, *Introduction to Group Analytic Psychotherapy*; Yalom, *Theorie und Praxis der Gruppenpsychotherapie*; Perry, *Couch Fiction*; Tuckman and Jensen, "Stages of Small-Group Development Revisited," p. 419; and Bion, *Erfahrungen in Gruppen und andere Schriften*.

26. Foulkes, *Introduction to Group Analytic Psychotherapy*.

27. Wilke, *The Art of Group Analysis in Organisations*, p. 3.

CHAPTER 7

ISSUES OF CULTURAL IDENTITY AND AUTHORSHIP WHEN RECEIVING TRAINING FROM OTHER CULTURES

Gražina Gudaitė

INTRODUCTION

The emergence of analytical psychology in Lithuania is closely related to the multi-dimensional process of psycho-social transformation which began after the country was liberated from the Soviet regime. The transition from an authoritarian system to a democratic one is a long process and includes profound changes in social life and in the life of individuals too. Soviet ideologies devalued individuality, creating an oppressive system against the diversity of the cultures of nations which were part of the Soviet Union. The use of the native language of a culture in official life was eliminated; the understanding of history was focused on the development of the Soviet state and the value of the mythology of these nations and their religious practices was very severely restricted. After the collapse of the Soviet Union, previously suppressed interests were legalized and became a flourishing aspect of group and individual life. This included an exploration of the history of Lithuania; involvement in religious life (practicing not only Catholicism, but also remembering old pagan rituals); and folk festivals and other activities showing the individual's need for deeper connections to their own culture and cultural identity. Joining the European Union opened further possibilities for developing individual potential, but questions of cultural identity today remain still not fully addressed. What is my place in the dynamic world and where do I belong? Who am I and how do I relate to others? These questions, of course, mirror those emerging within the analytic relationship.

Hypotheses about the Self and individuation, about efforts to understand the process of transformation and the individual's role in it, about connections between psychic life and culture and religion, and about the value of the symbolic all make analytical psychology very attractive in this period of transition. Introductory seminars led by Jungian analysts, the reading of C. G. Jung's texts and post-Jungian writings, and participation in conferences about analytical psychology all provide a rather different understanding of the inner life of individuals and generate the inspiration to look for ways to integrate analytical psychology into Lithuanian psychotherapy and culture. Professional training and the wish to become Jungian analysts and join the international community of Jungian analysts has been an important wish and hope. But after the collapse of the Soviet Union, to become a professional analyst was a mountain to climb as there were no professional Jungian analysts in the whole of Eastern Europe.

Models of Training

History shows that joining a relevant study program in an already existing accredited Institute of Analytical Psychology is one way to train as an analytical psychologist. This possibility was hard to implement for us because of socioeconomic differences. The costs of training programs were far too high for Lithuanians and also for people from other Eastern Europe countries. Even with some financial support (and some individuals did find money), the problem remains that living/studying out of your own country for a necessary period of at least five years and then returning home during a period of rapid change in Eastern Europe is bound to be very difficult. Individuals prefer to train in the country in which they live. Delivering a program of training from a foreign Institute to Lithuania was another possibility. This was tried in Lithuania, but, regretfully, the program offered provided no provision for regular personal analysis or regular clinical supervision, both the foundation for any training.

The third possibility was to seek out training through the International Association for Analytical Psychology (IAAP) Router Program, leading to individual membership of the IAAP. Such a model integrates theoretical study and personal work within a country of residence, as well as encouraging individual opportunities to train, including personal analysis and individual supervision with analysts from Western countries. This model of training permits students to remain in relationship with his or her own country, culture, and clients and takes account of the absence of analysts in students' countries. The IAAP takes responsibility for the preparation of training, organizing screening interviews, and examinations, which are part of the training regulation to ensure high standards of professional and clinical work. The IAAP also takes some responsibility to help establish relationships with suitable IAAP analysts for both personal analysis and supervision. This model was the most effective for Lithuania and has permitted us over time to form a group of internationally recognized Lithuanian Jungian analysts. We have applied successfully to become an IAAP Society and (at the time of writing) we are applying to become an IAAP recognized Lithuanian Training Society. Training within the frame of the Router Program was "good enough" during the transitional time of our culture and took into account our difficult economic situation.

Over the last two decades, twelve individuals have successfully passed through the Router Program and become internationally recognized Jungian analysts; others are well on their way. Besides training seminars in Lithuania, which were led by visiting IAAP and later by analysts from the newly formed Lithuanian Association of Analytical Psychology (LAAP), some members of the program found opportunities to train in Zürich, Los Angeles, Tel Aviv, and Chicago. Our trainees also had opportunities for regular individual supervision in Freiburg, Frankfurt, Zürich, Stuttgart, Berlin, and St. Petersburg, as well as attending seminars outside Vilnius, initiated by the students themselves. Training within the frame of the Router Program combined activities both in Lithuania as well as in foreign countries. Most of the client work and its supervision, participation in seminars led by visiting analysts, theoretical study, and much personal analysis or supervisions took place in Lithuania. For those candidates who travelled out of the country, it was mainly to supplement their personal analysis and for clinical supervision.

We currently hear different views about the Router Program. Experience has shown that it gives opportunities for individuals (who for many reasons are not able to join Institute programs overseas) to train and it contributes to the emergence of analytical psychology in different regions of the world. On the other hand, such programs have their limitations as the training structure is not strictly defined, the training regulations are rather diverse, the training process can be seen as lacking in continuity, and the evaluation of analytic competencies is sometimes complicated. There are many questions that arise and it is not easy to find common answers because of the diversity of the experience of trainees and their particular cultural-historical backgrounds and also because of the different and multi-dimensional processes involved in the relationships between candidates and their teachers.

Experience in the Router Program

In this chapter, I would like to discuss my experiences both of being a candidate and also of being a teacher in the Router Program. Some of my own studies in my training program took place in Chicago and other parts in Zürich. I have been analyzed and supervised by analysts from Chicago, Switzerland, and Germany. As a qualified Jungian analyst, I now work with individuals from Lithuania and from several other countries in Eastern Europe including Latvia, Estonia, Belarus, and Russia.

The structure of individual training following the model of the Router Program is flexible, and undertaking it depends on many factors—both internal and external. These include the candidate's personal motivation and training wishes, the opportunity to find analysts who agree to work with routers paying a low fee (which reflects large differences in earning capacities between Western and Eastern Europe), and willingness on the part of both trainees and their analysts to hold sessions on weekends, etc. Some social and political circumstances may also influence the training process, for example, getting a visa to study in another country or finding resources to be able to travel to another country. There are conditions too which may interrupt the process, but our experience has shown that maintaining continuity even in the most difficult of conditions can be possible. It seems to us that it was the quality of the relationships which developed between us and our colleagues from other countries which played a crucial role in building an effective and rewarding training process. It was not only the analytical relationships (which of course were the most important in personal analysis and supervision), but also human relationships, which helped to contain frustrations and open new possibilities for further development. A deep interest in and openness to meeting people from other cultures and mutual respect and engagement with the training development process all helped to overcome many financial and political obstacles and to compensate for the lack of clear and definitive structures and processes.

Psychotherapeutic research shows that if a psychotherapist and client have a similar background and language, this may facilitate the process of the therapy. We presume that when both participants in the psychotherapy are from the same culture, the patient may be more open to speak about difficulties because he or she expects the therapist to understand.[1] If we are from the same culture, this probably means we encounter some of the same difficulties and use the same coping strategies or healing patterns. Experience of "sameness" is an important condition for the formation of the inner structures and inner relationships in the psyche of an individual, and the mechanism of identification plays an important role in such a process. But what if partners in the analytical processes have different histories and different cultural complexes? How can the search for individual cultural identity be facilitated by a foreigner? There are complications to be encountered and questions to be asked within the field of relationships when people from different cultures meet.

All the analysts who trained through the Router Program in Lithuania reflected later that at the beginning it was difficult to speak in English. It was a frustrating experience (for some candidates English was their third language), but nobody mentioned that such limitations in verbal expression made the analytical process less rich. Almost everybody experienced the meeting with an analyst from another culture as a very special experience and commented that language did not interfere with the quality of the relationship which was the basis for an analytical experience. Words are important but something that stands beyond words moved the process on too. Our practice shows that the input of the visiting analyst depended not only on the information he or she brought to their lecture or seminar, but rather it was the style of presentation and the way he or she related to the students that played a very important role in understanding the ideas of analytical psychology. Openness and a deep respect for differences of opinion were paramount. This included interest in looking for a variety of manifestations of the psyche and real efforts to understand the meaning of unconscious communications. It involved the multidimensional understanding of the motivation and the phenomenology of experiences. All this was a new experience for us, especially in the context of our own confrontation with the consequences of an authoritarian regime. Meeting representatives from other cultures and integrating new models of behavior is a dynamic process and includes different moments of emotional experience which may veer from anxiety to surprise, from confusion to joy, from helplessness to hope.

Meeting the Other: Confusion and Discovery

Soviet people had become rather isolated from Western countries and following the breakup of the Soviet Union the opportunity to meet an analyst from Switzerland, Germany, or the USA, and to work together on a regular basis, was exciting but also a great challenge. Attitudes towards the West and expectations about relationships with people from other countries inevitably involved projections. Life in the West could be associated with paradise and going to the West was a real privilege. In the Soviet Union, only special people had such a privilege. In our experience, meetings with analysts from Western cultures and our expectations of them have been colored by parental complexes—giving and receiving, the need for mirroring and feedback, sensitivity to and understanding of the need to compete. These and other themes rooted in parental complexes were present in our work with the visiting analysts. On the other hand, meeting somebody from another culture includes opening the gateway to the realm of Shadow rooted in cultural unconsciousness. When I myself was being analyzed and supervised by analysts from other cultures, I was really surprised that so much of what we talked about related to Lithuanian culture and history (not only about my own or my clients' personal lives).

Perhaps the most surprising experience was that such sessions brought really deep relief. I remember how profoundly I was touched after sharing stories about my ancestors who died in exile in Siberia and hearing the hypothesis of cultural trauma and its transgenerational aspects in a person's life. Unconscious experiences were named and I could witness that finding that right name (as in a fairy tale) had a healing effect. The exploration of culture was important not only for a deeper understanding about suffering, but also in opening up new resources from these old stories.

I remember how members of our group were excited when the visiting analyst was giving a lecture about the Lithuanian artist Orvydas. In Soviet times, Vilius Orvydas created a museum, hosting exhibits of modern, sacral, and folk art. The exhibition was not accepted by Soviet administrators and was even destroyed a few times. It was a refuge for people who were not able to adapt to Soviet life—those who had to survive many of life's hardships. It was open for people released from prison and other dissidents. There are several pieces of art in the museum which symbolize the experiences of cultural trauma. Vilius Orvydas was not overwhelmed by the battle against Soviet oppression. He expressed a much wider perspective of the world, making sculptures which symbolized different experiences and different religions. This is one of the reasons why his sculpture garden evokes deep interest and generates a feeling of harmony not only for Lithuanians but also for people from other countries. Even during Soviet times, Vilius Orvydas accepted visitors from all over the Soviet Union, including Russia.

How may we understand the visiting analyst's decision to give a lecture based on his impressions after visiting the Orvydas museum? We knew the story of Vilius Orvydas ourselves and some of us had met him when he was alive. But do we remember all the stories we heard in our lives? After the breakup of the Soviet Union and after having open access to the information about cultural trauma (the huge numbers of people killed and tortured, numbers of Lithuanians exiled in Siberia, in prisons, etc.), the call for justice and the temptation to identify ourselves with the victim position at a cultural level was strong. The story of Vilius Orvydas suggests another kind of identification. The visiting analyst reminded us of our own story which contains experiences of loss, but it is not only about loss. The story shows that individuals may retain a wider perspective and can be open to different experiences with free choices even in very complicated historical circumstances. Of course, the price of the choice for freedom may be very high (Vilius Orvydas sacrificed a lot), but still the story gave us hope. Finding hope in our own culture can be a very important moment as it inspires us to look for personal resources to improve our self-esteem and self-worth in our own land.

Cultural Complexes and the Search for Identity

Our clinical practice in Lithuania shows that an individual is sensitive to his or her cultural background and cultural identity. As one client expressed it, "It is something, I can't change. And if it is related to shame, guilt, or fear, it is forever." Is this really so? To be Lithuanian, Latvian, or Russian connotes identity which may be rooted in some cultural complex. Naming someone by his or her nationality can trigger the complex which can evoke a defensive attitude and transference-countertransference dynamics influenced by cultural issues. For example, "I (as a Jew) do not know what it means to trust; maybe you Lithuanians know." This was said to me in a first meeting with a client whose family had been severely traumatized during World War II. Such a reaction was rooted in our different cultural historical backgrounds, rather than the personal. But we needed to face "otherness" and to work through it in our personal meeting. As psychotherapists, we may assume that when a client and therapist are from different cultures, the transference and countertransference dynamics are likely to be shaped by cultural values, and we do well to watch for a variety of reactions ranging from over-compliance and friendliness on the one hand to suspiciousness and hostility on the other.[1] We need to remain sensitive to questions of cultural identity and work on these in the analytic relationship, especially given the broad range of different cultural partnerships we encounter in the Router Programs.

Belonging to the same group and experiencing similar patterns of "being" and "doing" is important to generate a feeling of security, to find a feeling of being safe in the world. Our common culture gives us roles and models of interactions. We share the same myths and rituals when we face critical moments during our lives. This is part of a cultural identity. In Verena Kast's words, "culture can give us a map of meaning, incorporated in important texts, in religion, in shared symbols, in behavior which is understood by itself."[2]

Every identity (cultural identity too) has its shadow. When participants in an analytic journey are from different cultures, shame, guilt, or anger rooted in a cultural complex may be triggered. A pattern of victim and aggressor could be at the base of such a complex—envy too for perhaps having a better life, competition for who is the best, etc. These are just a few examples of particular dynamics that may become part of the analytical experience when we meet and work together with someone from another culture. The relationship may trigger the cultural complex and evoke projections rooted in it. Analysis of projections and understanding cultural complexes provides a good opportunity to develop a more realistic sense of cultural identity.

Cultural identity then may be defined as a feeling of belonging to some national group which becomes part of an individual's self-image based on experiences of cultural "sameness." A feeling of belonging depends not only on social history or the place where we were born; rather, a sense of belonging is about relationship, and it is a dynamic process, never completed. In therapy, awareness of the dynamics of cultural identity and hypotheses about the possibility of transformation can bring real relief. An individual is not able to choose his nationality, history, or culture, but then cultural identity is not just given. It is a subjective experience and there is a consciousness about it. Cultural identity does not have a "fixed origin to which we can make some final and absolute return. It is always constructed through memory, fantasy, narrative and myth. Identities are always relational and incomplete, in process if you like. It belongs to the future as much to the past."[3] That is part of the inner work of an individual living in transitional times.

Analysts who taught and supervised in the Router Program were very sensitive to our culture. Lithuanian myths and fairy tales and social history were prominent in seminars and private conversations. On the other hand, we were hungry to discover more of the world unknown to us so that we could discover the "otherness" of cultures different to our own, and our teachers were open to that too. Members of our group recount many special moments of visiting and experiencing other cultures: supervision and a summer camp in the Alps or Freiburg, experiences of Egyptian mythology in Luxor, having dinner with colleagues in an Afro-American restaurant in Chicago, a visit to the Getty Center in Los Angeles, etc. These and many other moments were privileges for us as they were available. Such experiences can be a very helpful antidote to the envy or longing for a better life rooted in our cultural complexes.

Combining friendship and analysis is a subtle process when the visiting analysts come to us. We organized our activities with a deep respect for boundaries which are so important in analysis. We set up clear structures for teaching and supervision during our meetings with visiting analysts. We tried to retain an awareness of group dynamics, and, when planning training activities, a respect for the privacy of all the participants in the process. Mutual interest in the differences between our cultures, openness, and a deep respect for "otherness" set the scene for a long-term relationship between colleagues from different cultures.

The Power of Internal Reality during the Training Journey

Within this chapter, I wish to emphasize the importance of the motivation of the individual who wants to train. By this I mean his or her feeling of a calling or inner readiness to begin the journey towards becoming a Jungian analyst. I would like to share a dream which illustrates the inner dynamics around the question of motivation. It is the dream of a candidate who planned to begin the Router Program, and, in order to do this, needed to leave her country for quite a long time.

> *I was appearing at a crossroads: the direction to the right was a well-built wide road for cars; the direction to the left was narrow, with grass and a muddy road for animals. Suddenly I saw an old and very big Japanese woman in the sky dressed in a kimono ... she said, "the direction to the right is the road of all people, go to the left." I was not pleased to hear such a command (especially when the road did not look easy), but I turned to the left.*
>
> *I was sitting near a little table in the corner of big sunny room ... A group of people from the West were sitting around a big table and making comments about my journey. I was listening silently to their voices while at the same time making some notes on the*

paper which appeared to be on a little table. To my surprise, the notes became colorful and I could see a sunflower. Then, it changed on its own and became three dimensional (like kind of sculpture). It looked like a really nice work of art.

This was a big dream, and it came back many times in different forms during personal analysis and individual training. The first item which attracted attention was the Japanese woman (associated to geisha), which in terms of the dreamer "appeared as a real stranger in the meadows of Lithuania." Meeting a stranger sounds similar as a motif when meeting an analyst from another culture. "We do not know who he/she is and we do not know what he or she brings to our life. He/she appears to bring change which can be both fascinating and frightening."[4]

Living in Soviet times, we had no contact with Japanese people, neither did we know much about their culture. Japan is very far away from Lithuania and nobody went to Japan to seek out professional contacts. The official Soviet ideology declared that Japan was an imperialistic country with classical differences and geishas were women with a bad reputation who were also victims of the patriarchy. The dreamer knew some Japanese poetry and she liked especially poems from the twelfth century. She remembered she was reading a memoire about geishas and she said she was impressed by the deep aesthetic sense expressed within it as well as its laconic style.

The figure in the dream was very big, showing herself in the sky; she was giving commands and seemed to have real power. This power came from above, a characteristic suggesting the phenomenology of a relationship with authority (which was real for the dreamer). This is a big issue for people living in a time of trying to integrate the consequences of an authoritarian regime. Psychotherapy research in Lithuania has shown the importance of this motif in the life of post-Soviet individuals. Authority was accepted as a dangerous, destructive, and defensive force and was a common theme in the psyche of clients who started psychotherapy.[5] Authority implies power, forcing others to take up a lesser, obedient position. In authoritarian regimes, the position of being humble was a necessary survival mechanism for the individual, and the projection of authority onto others is a common aspect of *Homo sovieticus*. In such circumstances, the capacity to internalize positive authority, especially in the form of a personal inner sense of authority, is inevitably rather limited. Experiences of low self-confidence, a defensive attitude, blocked creativity, etc., are all expressions of the so-called authority complex which was part of the psyche of *Homo sovieticus*. Our research has also shown that authority figures were mainly male figures who had power and who were often perceived as dangerous.

What message about the relationship with authority was conveyed in this dream? What kind of authority figures were Japanese women/geishas? Exploration of possible amplifications of this symbol revealed significant themes for the trainee. A geisha may symbolize an "all accepting" attitude as opposed to the defensive attitude of *Homo sovieticus*. It can be a helpful guide associated with Eros rather than the power principle. In Soviet times, the understanding of power was rather narrow and patriarchal including the devaluation of feminine energy. This dream shows the need and readiness of the candidate for another kind of authority. I know from personal experience, and within my clinical practice, that the search for another kind of authority is very common with clients from post-authoritarian regimes. Analysis of many dreams and personal memories has shown that liberation from an imprisoning system, a feeling of freedom, and the possibility of finding one's own individual path has been the most important attitude in this transitional period. A rehabilitation of the capacity to experience inner authority, integrating the feminine aspects of the personality, as well as the masculine, has been a common need in different fields of our lives. Such integration is a long process, sometimes successful and sometimes not, but, in such a context, the possibility of becoming an analyst within the framework of individuation and personal wishes suited Lithuanian professionals wishing to become analysts.

The dream does not say much about the nature of the journey, except it provides a warning that the road to the left is narrow but perhaps closer to nature with mud and grass. We can imagine that such a road or such a way of travelling does not provide a clear structure (as compared to the well-built road for all). It is not very comfortable, but rather symbolizes a journey that has to be experienced if there is to be progress. We do not know much about the journey itself, but the second part of the dream tells us what happens next.

Authority figures remain in the dream in the role of evaluators (one of the important functions of authority). Exams and evaluation procedures in all training programs will evoke many feelings and constellate complexes as well. Our experience shows that for some candidates, such an experience in the Router Program can be crucial. We can observe whether an individual needs a very clear training structure or if he or she is adaptable to the training requirements, including evaluations. If a candidate is open enough to the process and has good enough internal structures, then all evaluation procedures organized by IAAP are accepted as necessary hurdles to negotiate, but not the most important aspects of their training.

We see in the dream that authority figures take up a central position for some time, but later the focus of energy is concentrated on a creative act: the sunflower which grows into a nice piece of art *on its own*. The symbol of the sunflower may hold different meanings as it moves according to the position of the sun. We can also feel the importance of the aesthetic principle in this moment, which is a good connection to the first part of the dream. There are many other aspects of the symbol of sunflower, but I would like to highlight here the fact that the transformation in the dream happens *on its own*.

The second part of the dream illustrates a bigger picture. There is a group of people who help to reflect on the process and who evaluate it; and there is a dream ego open to listening to other opinions, but at the same time acknowledging that the flow of energy and change happens *on its own*. Some things happen spontaneously without conscious readiness or foresight. This might be described as an "action or manifestation of the self" linked to the power of internal reality.

Conclusions

Who is the author of the analytical process or, indeed, of the analytical training? If there is someone (an institution or a group of people) who is willing to deliver a program and others who are open to receive it, then there is potential for a partnership. In our experience, questions of ownership and authority (who regulates and gives structure) are rather complex and depend on many factors which vary with time. At some stages of training, it may be only one analyst; at other moments, the group is the author, including the group that is the IAAP who writes the guidelines and procedures. Sometimes candidates' decisions provide the main energy which moves the process on and gives structure. Perhaps the most important transformational energy happens between people, creating a new "third" from the depths of that relationship. In Cwik's words,

> The nature and experience of this unconscious third during analysis directs the process at any given moment and is shaped by underlying archetypal patterns. This in turn leads to being able to live a more fully symbolic life—a life lived with sense of aliveness imbued with meaning and purpose.[7]

An analytical experience contributes to the capacity to live a symbolic life which could mean a deeper connection to culture as well. An awareness of one's own cultural roots and mores, the experience of being the same as others, being in touch with a stranger, and, finally, the awareness of difference, all help each of us to develop a realistic sense of identity.

By the way, the word stranger in Latin is *hostis,* which can mean both "guest" and "host" and is the root of the word *hospitality*. Meeting analysts from other cultures through the Router Program gave us an experience of both being a host and offering hospitality to analysts who came from other cultures, and also of being a guest visiting other countries with many opportunities to explore and experience the culture of the so called "other," who may have been perceived as strangers at the beginning, but who later became our colleagues and our friends.

Such communication is a non-traditional structure for an analytical training, but it has worked in Lithuania. Perhaps there are some limitations to working with the transference with such a model when we need to travel to the important figure or wait patiently for his or her arrival. This may indeed be an obstacle, especially when working with early and more primitive processes and especially in the transference where complexes are rooted at pre-oedipal levels of development. It is more difficult too for candidates who need to do analytic work to develop a more secure ego structure. Although the lack of continuity and possible fragmentation of the training process points to dangers for some candidates, this is life! Even when we explore the phenomenology of the formation of cultural identity, we find comments that identity is framed by two axes or vectors, simultaneously operative: the vector of similarity and continuity; the vector of difference and rupture. ... [T]he one gives us some grounding, some continuity with the past. The second reminds us that we share precisely experience of a profound discontinuity.[8]

Participation in the Router Program is a journey, different in character for everyone, and the transference remains important as it helps to identify the helpful forces, the messengers, and other unconscious forces.

Being the same and being different to others—meeting those who are similar and those who are different to us—is like being alone and being together. Both are part of relating and individuating.

We thank all the analysts who came to Lithuania and shared their knowledge and experience with us and who accepted our hospitality. We thank all the analysts who opened the doors of their houses and rooms to us and who shared their hospitality and culture. "If a man be gracious and courteous to strangers, it shows he is a citizen of the world and that his heart is not an island cut off from other islands, but a continent that joins them."[9]

Notes

1. Carolyn Rodriquez et al., "The Role of Culture in Psychodynamic Psychotherapy: Parallel Process Resulting from Cultural Similarities between Patient and Therapist," *American Journal of Psychiatry* 165 (11, 2008): 1402–06.

2. Lillian Comas-Diaz and Frederik Jacobsen, "Ethnocultural Transference and Countertransference in the Therapeutic Dyad," *American Journal Orthopsychiatry* 61 (3, 1991): 61.

3. Verena Kast, "The Cultural Unconscious and the Roots of Identity" (keynote address, First European Conference of Analytical Psychology, "Dialogue at the Threshold between East and West: Cultural Identity Past, Present, and Future," Vilnius, Jun. 25–27, 2009).

4. Stuart Hall, "Cultural Identity and Diaspora," in T. Rutherford, ed., *Identity: Community, Culture, Difference* (London: Lawrence and Wishart, 1990), p. 226.

5. A. Ronnberg and K. Martin, eds., "Stranger," in *The Book of Symbols: Reflections on Archetypal Images* (New York: Taschen, 2010), p. 486.

6. G. Gudaitė et al., "Revelation of Personality's Integrity: Problem of Relation with Authority and Psychotherapy" (research, 2014–2016); this research is funded by a grant (No. MIP-080/2014) from the Research Council of Lithuania.

7. August Cwik, "From Frame through Holding to Container," in Murray Stein, ed., *Jungian Psychoanalysis: Working in the Spirit of C. G. Jung* (Chicago, IL: Carus Publishing Company, 2010), p. 169.

8. *Ibid.*

9. Ronnberg and Martin, "Stranger," p. 486.

Chapter 8

Women and Professional Identity in Russia

The Doll of Vassilissa the Beautiful as a Symbol of the Handover of Feminine Knowledge from Mother to Daughter

Elena Volodina

Individuation as Container: The Search for Professional Identity in Russia

Natalia Alexandrova

The Doll of Vassilissa the Beautiful as a Symbol of the Handover of Feminine Knowledge from Mother to Daughter
by Elena Volodina

> You are a woman and thus you are right.
> —Valery Bryusov, *Stikhi* [Poems]

A woman and a profession: How are these two notions connected? How can one remain a woman while being a professional, and become a professional without letting this affect her femininity? What does being a woman contribute to the job of a Jungian analyst? Many patients specifically request a female analyst.

What are they after? Is there a connection between becoming a woman and becoming a professional analyst?

In 1998 the International Association for Analytical Psychology (IAAP) launched its first project of training analytical psychologists in Russia. It set up a training program of lectures, workshops, and supervisions aimed at helping Russian analysts achieve the professional standard of international membership in the IAAP. I took part in this program in Moscow as a trainee from 2006 to 2010 when I qualified. In the framework of that training program we worked with supervisors and analysts from the UK who regularly flew to Russia for this purpose. Most of them were female.

The Russian cult of sacred femininity did not include either the sanctification of purity or the worship of beauty. This is why in Russia, Mother Mary is not called either the Holy Virgin or the Queen of Heaven, but the Godmother, which means that motherhood is perceived as her key quality. In old Russian spirituality, eroticism is less important than work, which means that in Russia "Demeter is described as stronger than Aphrodite."[1] In Russian culture, it is the motherhood of God's Mother that is worshipped. The woman had always been protected in Russia as the one who enables the continuity of generations. The archetype of the Russian woman had always been in essence that of a mother. "The woman is potentially the mother not just of separate beings but of nature as a whole and of the entire world, fallen and helpless in its downfall."[2]

In Old Russia women lived in compliance with the moon cycle, keenly responsive to signs from nature. They gathered medicinal herbs and knew how to heal the body and cure the soul. Almost any woman was a healer in her family. And in every village there was a *vedunya* (literally "the knowing female"). A *vedunya* had an innate gift of seeing the ties between the spiritual and the physical worlds. She possessed ancient knowledge, clairvoyance, and spiritual sight. The Slavonic *vedunya* was a figure of authority, a wise woman, the spiritual advisor of the community, and the key source of advice on all matters of daily life and family. Her very presence contributed to the prosperity, health, and spiritual balance of the community. The word *vedunya* comes from a root which means "to know, to possess knowledge." The secrets of her knowledge were passed on from mother to daughter, from grandmother to granddaughter.

In the Russian fairy tale "Vassilissa the Beautiful," a dying mother presents her daughter with a doll, made with her own hands, which helps her on her path to adulthood and the bloom of her femininity. It helps the development of a woman's wit, strengthens the feminine, and grants the courage to navigate the outside world with confidence. That wisdom had been handed over for centuries from mother to daughter, strengthened and nurtured to be passed on to the next generations. Vassilissa's doll is the symbol of a mother giving her daughter her sacred understanding of feminine nature: the knowledge of how the female organism functions in unison with the moon cycle, acceptance of the miraculous birth of a new life inside a woman, the capacity to bear and raise children, and to do it with love and in harmony with nature. The doll given by her mother helped the woman to open up her soul, to discover her destiny; it allowed her to be loving and charitable.

* * *

It had always been like that, and it seemed that it would always remain so, however, in the beginning of the twentieth century, time went out of joint; there was a revolution. The old traditional life was over, new times began.

While in the nineteenth century, praise was sung to the masculinity of a woman (the poet Nikolay Nekrasov admired the Russian woman for being able "to stop a galloping horse and enter a burning house"), the twentieth century dealt the feminine a heavy blow.[3]

As Hermann Hesse says, "Human life becomes a real hell of suffering only when two ages, two cultures and religions overlap."[4] In the twentieth century, the people of Russia were forced to survive in exactly such an overlap. Several generations were so tormented by the sickness of their time that they stopped acting according to the cycles of nature and broke the continuity of tradition, losing any protection they had.

So many men died in the revolution, wars, and Stalinist purges, that women had to take their places. All efforts were directed at survival. The main objective of society was to build communism, which meant that work was the most important thing. Family values sank into the background. Marriages were as easily entered into as they were dissolved. Everybody lived publicly: in hostels or communal apartments. One lived one's private life in the public domain, which meant that one had to hide, contrive, and lie. A lot of shame was involved.

Pregnancy did not bring happiness; it was a complication that had to be dealt with. The first thing a doctor would suggest was to terminate the pregnancy. Abortions became a common thing. Many women had dozens of abortions in their lifetime. Pregnancy was an ordeal that involved queuing for hours for maternity appointments and getting tested for syphilis. Doctors did their best to record a date as late as possible for the delivery, so that the mother would get as short a maternity leave as possible. Finally, women had to face the horror and humiliation of a maternity clinic, where the doctors were cynical and rude, and nappies were in short supply.

After childbirth, a woman could stay at home with her baby for a very short time. After just two months, she had to go back to work, while the baby was sent to a nursery, doomed to deprivation and early separation. The universal entry of children into kindergartens was interpreted as the liberation of women from the burden of housework and childcare in favor of creative labor. Mothers did not breastfeed their babies and did not hold them. All this led to the breakdown of the maternal instinct and contempt of the feminine function.

Mothers did not know how to tell their daughters about their painful and shameful experience, and preferred to say nothing. Most girls were not told by their mothers even about periods, let alone sex or childbirth. What should have become the initiation of a new life, a wonderful opportunity to use one's creative function to give birth to new life, became a source of shame and trauma. The handover of traditional knowledge from mother to daughter was disrupted. The destiny of a woman degenerated into being humiliated and servicing the production process by giving birth to workers, the builders of communism. It became a new production line. Life became a struggle for existence under the conditions of constant shortage of commodities and shortage of love. Tired mothers perceived their children as nuisances. Children spent more time in childcare institutions than in families. They had no concept of home. Such mothers had little to give their daughters except their absence and, of course, the absence of a doll.

The Russian woman can easily sacrifice anything for the man she loves and for her children. Yet the sacrifice of the feminine fate and love to social ideals and the building of communism was the greatest sacrifice made by women in the twentieth century.

Women lost touch with their internal woman, their natural femininity, and the spiritual component of a woman. Daughters did not understand what it meant to be a woman; they never passed any rite of initiation and did not get Vassilissa's doll. The handover of the feminine tradition from mother to daughter was disrupted. *Vedunyas* were gone from the villages and there was no one left to heal or offer support. The balance between the masculine and feminine was upset; the integrity of the world was lost.

As women were forced to become breadwinners, they grew more and more independent from men. The numbers of single women and single mothers grew. Society did not appreciate femininity, gentle or delicate feelings. Whatever made life special stood in the way of survival. The survivors were ones who preferred thinking to feeling and only used their intuition to sense danger. This is why nowadays many smart, highly educated women are entirely out of touch with their feelings. They can successfully survive in harsh conditions but do not know how to live.

* * *

A patient of mine—a young and attractive lady, intelligent and well educated, successful in her career, but lonely and childless—once had an image in one of her dreams:

> *A female body is divided into pieces. Morsels are packed into plastic bags and are pushed deeply under the sofa; they have been stored there for a long time already, covered with dust and dark.*

Horrified, the patient became aware that it was an image of her femininity, with which she was not acquainted.

To my mind this image is a very important one—it might be considered as a common image for a majority of women who are living in the post-Soviet space. It is the image of a collective trauma that was done to the feminine by all the wars, revolutions and ugly social experiments. Of course, femininity dwells somewhere, but it is hidden in a far distant place, in dust and darkness, and a certain effort is required to find this femininity. And then it will become obvious that we have first to liberate it from wrapping and only after that to face the seemingly unfeasible task—to join up the cut-off pieces, to help them to grow together and to revive—a task that is similar to a scary fairy tale.

This is a long and laborious work that we are doing in analysis. Due to the length of time—the lifetime of several generations—that this dismembered woman has been lying under the sofa, we need a long time for coming into awareness and for healing.

For most of my female patients, the need for a therapist represents the need for a good mother. A negative maternal complex is clearly present in their psyches. Almost all of them have passed through the furnace of nurseries and kindergartens. They are still haunted by rows of chamber pots, closets in which they were locked as a punishment, or a screaming monster of a nurse. Early separation and deprivation have prevented them from establishing harmonious communication with their children. They have to be taught to do it. These women, whose mothers had dozens of abortions, are tormented by the question "why didn't she kill *me?*" Their mothers did not tell them a child was a miracle and a joy; instead, they know that a baby is a nuisance, a hindrance to future life, and a source of suffering. Their Mother Archetype contains very few positive personal images. The key goal of Jungian therapy in such a case is enabling the patient to receive images of the Positive Mother, which will help her pass various feminine initiation rites, get in touch with her internal femininity, and acquire confidence in her feminine destiny, e.g., the doll of Vassilissa the Beautiful.

In the race for survival, we have squandered the feminine strength that nature gave us. Over time, we have lost the concept of female destiny and learned to value masculine qualities, thus contributing to the strength of the Animus and pushing femininity into the Shadow. For other professions, this may be unimportant or even beneficial, but for a female Jungian analyst, it is essential to be in touch with her feminine essence. We the female analysts who grew up in the Soviet Union have had the same experience as our female patients; we are carrying the consequences of the collective trauma, and thus we need a special aid and support to be able to help our patients. Just like them, we did not get a doll from our mothers. Our main goal is therefore to restore and heal our femininity.

During the IAAP training we were analyzed by qualified Jungians who came to Russia regularly four times a year. They were female analysts from the UK, a country where traditions are valued and families can be traced back several centuries. These women's life stories and female experiences were dramatically different from ours. In the course of analysis, the positive side of the Mother Archetype was filled with the images and energy of the British women who never knew Stalinist purges, the horrors of collectivization and industrialization, or the dread of kinship—of having to repudiate their relatives to avoid repressions. They did not just teach us the tricks of the trade. Consciously and unconsciously, they transferred to us natural feminine values and expressed horror at what we had endured. In this atmosphere of acceptance and support, our femininity was healed, it felt as if everything that was bent and broken was repaired, and the magic of a female healer came back to life. We were born again and could hear the voice of the internal ancient woman inside us. The British analysts acted as mothers that gave us their knowledge, traditions, and female secrets in order to mend the broken ties between generations and release the flow of energy. When our training was complete, the British analysts left each of us a symbolic doll: the secret knowledge of our creative feminine power.

The cultural overlap was highly beneficial to the training process; the traditions and images of a different culture enriched us as people and professionals. The organization that sponsored the training program was called "The Russian Revival Fund." The name referred to the revival of analytical psychology but what really happened was the revival of an ancient Russian tradition: the transfer of a *vedunya*'s skills of soul healing from mother to daughter. In the course of the training, a magical doll was created and handed over to us, the female Jungian analysts of Russia.

Individuation as Container:
The Search for Professional Identity in Russia
by Natalia Alexandrova

> The unfortunate generation of Soviet people all mutilated by the war—the generation of the armless, legless, burned and physically crippled, but living among plaster and bronze figures of workers with athletic arms and peasant women with strong legs—this generation despised any frailty.
> —Lyudmila Ulitskaya, *Sincerely Yours, Shurik*

In this chapter, which addresses the development of professional identity, I have attempted to link several issues which are important for me and through which I worked both in my personal analysis and in therapy practice with my clients. My reflections and my personal analysis have long revolved around experiencing and comprehending collective trauma and the way trauma is transferred to the next generations, children and grandchildren. Through this process I made an important discovery, an insight that revealed itself gradually—*intergenerational trauma is silent and is denied*. Silence and denial are two interconnected defense mechanisms used to cope with unbearable pain. I believe Dolto conveyed the same message when she said "what is silenced in the first generation is buried in the body in the second generation."[5] Silencing means not only what must not be spoken but also something which is not possible to speak about due to the painful experiences associated with the trauma. Jung wrote about what silencing by parents means for their children; "the children will have to suffer from the unlived life of their parents and ... they will be forced into fulfilling all the things the parents have repressed and kept unconscious."[6] The parents' silence confuses the children (the children both know and do not know about the trauma); the children in turn remain silent too, as they understand that they are not to ask questions—therefore, something unspeakable appears in the family. This mechanism also emerges in analysis—clients may complain that there is something in analysis that remains unrevealed. I believe that my interest in analytical psychology was awakened largely by its special focus on the mechanism of how the content of the unconscious is communicated from parents to children.

Pickering pays special attention to the intergenerational transmission of trauma through the dreams of children whose parents have experienced trauma:

> Untold and unresolved intergenerational trauma may be transmitted through unconscious channels of communication, manifesting in the dreams of descendants. Unwitting carriers for that which was too horrific for their ancestors to bear, descendants may enter analysis through an unconscious need to uncover past secrets, piece together ancestral histories before the keys to comprehending their terrible inheritance die with their forebears.[7]

Studying collective and transgenerational trauma also requires attention to the issue of the mix of the personal and the collective, an unconscious confusion of identities, and finding ways to seek the true self. Connolly, who has worked as an analyst in Russia, notes "the presence of intergenerational trauma not only in the children of Holocaust survivors but also in the children of the survivors of repressive regimes."[8] Millions of men and women—crippled, injured or killed in war, tortured in concentration camps, women left alone with small children—all comprise a cultural layer that gave birth to several generations of Soviet people.

Actually, my key question is about how the traumatic experiences of parents and mothers affect the child's emotional development in particular. It can be assumed that significant negative consequences of collective trauma impair the mother's ability to stay in reverie providing the secure container to her baby.[9] The mother's impaired ability to stay in reverie stimulates the child to find some method of self-identification in the surrounding world that could compensate for the absence or inadequate presence of a maternal container. Yet this missing experience is vital for the integrity and security of the child's physical and mental development.

According to Schellinski, the trauma affects the survivors' children and may precipitate identity and personality disorders, depression, or difficulties in relationships, which are linked to survivor's guilt, unaccomplished parental mourning, and the replacement of dead children and relatives. When a person has to take on the role of another who is to be replaced, this person is deprived of an opportunity to live his or her own life.

> From before the child is born, the archetypal forces of death and life are joined in a fateful constellation; the soul of the replacement child bears the shadow of death from the very beginning of life. Hope for the replacement child lies in an emergence of true self as soul recreates original life ... the path of individuation countering the replacement child's identification with the dead.[10]

CLINICAL EXAMPLE

I would like to present a clinical case where, I think, we can see the impact of the parents'—particularly the mother's—traumatic experiences on the inner world of the patient. This story was at times difficult to bear and analyze. In the process of analytical work the mother-child pair relationship was recreated through the mechanism of transference and countertransference. In my countertransference I experienced that when the mother container is destroyed there is a need to find some particular cultural container. I felt like an emotionally-devastated mother looking for something that would help me to survive. Now I can say that I was in search of a cultural container which could heal myself to some extent. Also, I was looking for a cultural container in which I could place the terribly scared inner child of my client. For me, Jung's concept of individuation became just such a cultural container, allowing me to move forward even at the most desperate moments. According to Feldman, Jung "has left us a remarkable legacy of a creative encounter with the depths of the human spirit. His work has shown how the inner world can compensate for a painful and tormented outer experience in a creative and transformative way."[11]

A man of thirty-five, whom I will call V, presented for treatment with severe depression. Since childhood V had been expected to show outstanding results in painting, where he failed. He had studied in an arts school for gifted children where it became clear that he was not talented enough to be a painter. Nevertheless, he entered a department of arts at university, graduated from it, but was unable to become a good artist and took his failure very hard. Then he tried his luck in business, but was unsuccessful again. He defined his life mission as "I must do something significant, something great;

I must achieve success or implement an important social project." He lived under the pressure of that mission and compared himself to Sisyphus rolling his boulder up a hill, watching it roll back down, only to do this over and over again.

V described his life as a succession of alternating states. In his active periods, he could think clearly, communicate with others, and make plans, while in depressive moods he stayed in bed, switched off the phone, and could not go out. He complained of an unbearable state of weakness and of being unable to accept himself in those periods. That weakness exacerbated his anxiety and fear of incapacity to control the situation. He attacked himself for being helpless, and called that part of himself "an idiot." He was haunted by two inner figures—a male and a female. The male figure was personified by a Man in Black, and the female by the goddess Kali. Both figures were aggressive towards him and craved for his slow and painful death.

In addition to having fantasies about his tormentors, he told me about a Trickster figure, whom he called the Joker. The Joker was cheerful and optimistic, capable of finding a solution to any difficulty. It was a supportive figure, so V could talk with him and get advice. A special feature of the Joker was that he possessed a superior physical ability, and had appeared when V was a child with numerous illnesses.

V's relationships with women were complicated and unstable, filled with hatred, revenge, and humiliation. V said that he was attracted to women who rejected him; he tried to win them, and when they seemed interested, he would reject them in response. Besides, he seduced and engaged in sex with women whom he regarded as inferior, using them only for sex and rejecting their affection. When we started sessions, V was not married and had just broken up with a woman after a long-term relationship—she had left him for his friend.

In the course of therapy, V spoke about his strong desire to change his name. He believed that personal history and self-identity are closely connected with the name a person is given by his or her parents. Like his uncle and namesake, V had been named after his grandfather. Both his grandfather and his uncle had died of lung cancer, and he expected to face the same destiny. V had strong self-destructive inclinations which made him unconsciously look for people and situations that would confirm his own sense of worthlessness and having no right to live.

Family Background

V was born in 1956, on the day when Russian tanks rolled into Hungary. In his mother's view, her son was born into a horrible world. She had graduated from a medical institute in 1941 and worked as a surgeon in a hospital through the Second World War. The hospital was flooded with patients, and her work demanded enormous physical and emotional efforts. She performed surgery on injured soldiers and officers; some of them survived and some died. Many of her friends were killed; her brother was killed in action and a man whom she loved dearly was killed too. So some part of her life ended with the war, but the psychological, emotional trauma that had affected her in the war has stayed with her for life.

The war had affected V's father too. A musician by profession, the father had been full of lofty ideals before the war. During the war he served in the military reserve forces, his ideals were obliterated, and he considered himself a coward. He developed a survivor's guilt—almost all the best people had been killed and he had survived despite being a loser.

V described his childhood as a life continuously controlled by his strict, authoritarian, and anxious mother. His mother's anxiety was mostly focused on V's physical state—she treated him for heart defect, annual pneumonias, pyelonephritis, and so on. He described his feelings towards his mother as a combination of fear and hatred. When the fear subsided, he would perceive their relationship as a world where he had to learn how to win, which meant that he had to learn to use people more efficiently than his mother could do.

He regarded his relationship with his father as more positive; they used to play together, but quite early V became aware that his father was weaker than his mother. Those childhood emotions had evolved into a grievance against his father: "You gave me no chance to respect you because you were weak; so now I am unable to have respect for men and, therefore, for myself."

I started therapy sessions with V at the beginning of my career as a therapist. I found working with him exciting; he inspired me with his energy. His life seemed fascinating, unusual, and full of events. I could easily sense his admiration for power and his contempt for weakness, and I was eager to help him. On the other hand, V sometimes provoked intense feelings of rejection, disgust, and contempt in me, and it seemed intolerable to work with him. I felt doomed to be tortured as I knew that he would humiliate me again and again, as if trying to destroy me.

Such polarities were frequent in our relationship. At times, I was deemed totally useless and inexperienced, and he poured endless streams of contempt, humiliation, and superiority over me. He claimed that I was a worthless and incompetent therapist, which fitted into a more general message: "I have to start a new life, replace old friends with new ones, have a new therapist and a new name." His major claim was that I was not able to solve his problem, to handle his depression. He showered me with his profound despair, and I felt paralyzed, unable to think or talk.

Then the picture would reverse, and I would turn into a wonderful and sensitive therapist (while those amazing qualities were declared to be the result of his influence—I had developed them under his guidance). So, the fact that I was a good therapist was not my achievement, a result of my studies or development, but it was to his credit because I had acquired my professional skills only through working with him. I was a "piece of clay" in a master's hand. In those moments V took a patronizing role and felt very important. In my countertransference, I envied him for having such a good therapist. Whose envy did I perceive? It seemed to be the envy of someone's carefully-arranged, successful life.

He had nightmares, and here is one of his recurrent dreams which V first had when he was four years old:

I am alone in a desert after a nuclear explosion. I am the only one who survived, all the other people have died. I don't know what to do; I feel helpless and hopeless.

The image of death frequently emerged in the therapy—in dreams and in active imagination. In V's associations the image of death was connected with the internal image of his mother who hates him and wants to kill him, with a lack of permission for independent life and love. There was an endless struggle between two sides—an abuser and a victim—in V's inner world. He projected that struggle into reality and into our relationship in the transference.

Gradually his attacks on me decreased. He described changes in himself using the language of body symptoms—he said he started to feel his heart. That signified a change in the therapy; he came to me to receive emotional warmth and could openly say it. I suppose we managed to create a third dimension through our therapy—one of kindness, warmth, compassion, and a possibility of living in this new space.

In the seventh year of the therapy V had a dream that was quite distinct from his other dreams, and V called it "different." In terms of individuation, I see this dream as an archetypal process full of archetypal images:

> *Bare rocks, dead vegetation. I feel anxious and doomed. There is a maze in a cave which pulls me inside. It is dark and wet, the walls are slimy. There is a stream flowing by. At first, I am moving along a tunnel, and then apartments begin to appear. I walk through each apartment and jump down to the next level of the maze. I am scared, and I do not want to jump, but I still want to move on. I feel elated after each jump. Gradually the cave becomes inhabited. Each apartment consists of a kitchen. There are wooden cupboards and tables in the style of the '60s. There are sprigs of herbs hung up. Sometimes I see linen hanging to dry; I catch various smells. It is cozy there. While passing by, I see men and women. In those apartments, I can feel life, something very casual, it's a place where I want to go to bed and live normally. In each next apartment the design becomes more exquisite. The last room is a veranda, where a writer could spend summers. One wall of the veranda is open and faces a seaside with a sandy beach. There are few furnishings on the veranda, but they have all been selected with great taste. There is a mahogany table and a white porcelain vase on it. The vase has a lid. There are three hieroglyphs and the English word "power" on the vase. I wish to open the lid, and at the same time I fear to do it.*

This dream occurred when our therapy was coming to an end. By that time, V had met a woman who was a painter. They started dating and then living together. Sometime later they had a son. Last time V phoned was when his son turned seven and entered an arts school.

Conclusion

While analyzing our work, I was looking for a context where I could place V's emotional experience, and I found it in works describing the children of Holocaust survivors.[12] The mother's wartime experience affected her capacity to sympathize with and be emotionally available for her child. V was born at the time of a new danger in Hungary, which created an additional re-traumatizing factor for his mother. V was born as a replacement child for his mother's family and friends whom she had lost. V was the child who was to compensate for his mother's irreparable losses, make her broken dreams come true, and embody her ideal projections. V's personal wishes or interests did not count. His mother's devaluation of his father did not allow V to identify with the father and receive support for his own masculinity. The above can explain V's strong wish to be reborn as a compelling need to be separated from his mother's introjections and to acquire a life of his own.

This case is a description of deeply disturbed motherhood and the subsequent results of it. Analyzing this case I began to think about the transgenerational trauma and its effects on the second and third generations. In the analytic relationship through the processes of transference and countertransference we reconstruct the life story of the analysand. What kind of a container for this person can I provide in my work? How much of this unbearable experience can I hold and give meaning to? Reflecting upon my professional choice and searching for my professional identity I can say that my trust in Jung's idea of individuation became my cultural container which allowed me to recognize and find meaning in this work with the effects of transgenerational trauma.

Notes

1. Georgy Fedotov, *Sud'ba i grekhi Rossii* [The fate and sins of Russia] (Moscow, Dar, 2005), p. 5.

2. Nikolai Berdyaev, *Filosofiya tvorchestva, kul'tury i iskusstva* [Philosophy of creation, culture, and art], vol. 2 (Moscow: 1994), p. 300.

3. Nikolay Nekrasov, *Stikhi* [Poems] (Moscow: Khudozhestvennaya literatura, 1965), p. 85.

4. Hermann Hesse, *Steppenwolf*, trans. David Horrocks (Penguin Books, 2012), p. 12.

5. Quoted in Anne Ancelin Schützenberger, *Sindrom predkov. Transgeneratsionnyye svyazi, semeynyye tayny, sindrom godovshchiny, peredacha travm i prakticheskoye ispol'zovaniye genosotsiogrammy* [Syndrome ancestors: Transgenerational communication, family secrets, the anniversary of the syndrome, the transfer of injuries, and the practical use of genosotsiogrammy] (Izd-vo Instituta psikhoterapii, 2001), p. 30.

6. C. G. Jung, "Analytical Psychology and Education: Lecture One" (1924/1954), in *The Collected Works of C. G. Jung*, vol. 17 (London: Routledge and Kegan Paul), § 154.

7. Judith Pickering, "Bearing the Unbearable: Ancestral Transmission through Dreams and Moving Metaphors in the Analytic Field," *Journal of Analytical Psychology* 57 (5, 2012): p. 576.

8. Angela Connolly, "Healing the Wounds of Our Fathers: Intergenerational Trauma, Memory, Symbolization and Narrative," *Journal of Analytical Psychology* 56 (5, 2011): 607–26.

9. Catherine Crowther et al., "Fifteen Minute Stories about Training," *Journal of Analytical Psychology* 56 (5, 2011): 627–52.

10. Kristina Schellinski, "Who Am I?," *Journal of Analytical Psychology* 59 (2, 2014): 189–210.

11. Brian Feldman, "Jung's Infancy and Childhood and Its Influence upon the Development of Analytical Psychology," *Journal of Analytical Psychology* 37 (3, 1992): 255–74.

12. Angela Connolly, "Healing the Wounds," p. 610.

CHAPTER 9

INFLUENCED, CHANGED, OR TRANSFORMED?
REFLECTIONS ON MOMENTS OF MEETING IN A BORDERLAND

Tomasz J. Jasiński and Malgorzata Kalinowska

It has to be faced that borderlands exert a dynamic, inspiring and optimizing effect upon all situations, possibilities and emotions that are happening inside of them. ... Everyday experiences of diversity in customs, languages, religions and even products of daily use, impel certain choices, judgments, adaptations and fascinations. And this leads toward independent thinking, creative evaluation, and acceptance of otherness. The mind and the imagination of the people of the borderland are thus in constant movement to which the monotony of thought and indolent expediency, so representative for the members of the homogenous areas in civilization, is foreign. ... But this adventure will not last for long. Because the days of borderlands are numbered. The world is becoming a cultural monolith, a sullen phantom that is feeding us with soap operas and advertising of global supermarkets. Therefore the spirit of the borderland becomes hidden. It went into exile. It came down into

the underground, ashamed with its own anachronism.
It doesn't disclose itself and lives in concealment. But
it goes on living and has a lot to say.
—Zdzisław Skrok, *Wielkie Rozdroże: Ćwiczenia Terenowe z Archeologii Wyobraźni*

INTRODUCTION

"Just as there is nothing more unchristian than the history of the Catholic Church to a Protestant, so there is nothing more unanalytic than the institutional history of analysis to an analytical psychologist." So says James Astor in the beginning of his paper on supervision and training.[1] With this provocative statement, he sets a new standard for those who write about the dynamics of Jungian training; a standard that will aspire to embrace both the transformative character of the training as well as its shadow aspects. Any discussion of training brings to mind a number of different areas of interest: the relationship between different training institutions within the field, trainees' attitudes to institutional dynamics, differences between training programs, the context of an individual transformation, dangers embedded in training, the relationship of training to analytic identity, the role of supervision and personal analysis in a training, etc. All these areas are important. We would like to add one more which is the role of culture and transcultural factors in training.

Interestingly, these two factors of institutionalization and transcultural influences seem to be intrinsically intertwined, because the latter underlies the very beginnings of analytical psychology. The transition from a movement that emerged from the work of one man to institutional organizations that make up the Jungian world of today took place through multicultural exchanges. The process of transition from peripheries to social institutionalization gained wider recognition, new status, and legitimacy. As it brought the fruits of structured institutes and societies, we once again faced, to use Homans' metaphor, the transcultural dynamic with its borderland-like permeability of multicultural exchanges.[2] This was brought on by political and social change at the end of the twentieth century—the fall of communism and the Iron Curtain and the opening of exchanges between geographically distant countries as a result of globalization.

These factors have affected psychoanalytic training. Within the Jungian world they also resulted in the formation of the International Association for Analytical Psychology (IAAP) Router Program. This form of training evolved in areas of the world where there were no Jungian analysts. Any training activity then, be it theoretical seminars, personal analysis, or clinical supervision, involved meetings between trainers and trainees from different cultural spaces.

Within the last twenty years, the IAAP Router Programs have become the fastest growing Jungian trainings in the world. From small beginnings, they brought a new life to Jungian theory and practice.[3] But they also raised concerns. Within the community they have been a source of passion, satisfaction, confusion, and critique. They awakened a divergence of opinions from enthusiasm and seeing them as "the hope" of analytical psychology to critical views that would see them as trainings producing "lesser analysts" who have not gone through "good enough" preparation for becoming analysts. In our opinion most discussions so far about the router training have focused on standards and dynamics, including troublesome processes within so-called Router Groups. But they failed to address two basic questions: what does this kind of training tell us about the situation in the Jungian world in general?[4] Then, even more importantly, does the router training provide the psychological space at both an individual and group level for a deep process of transformation?

The Router Program is in itself a particular form of training. It doesn't have a typical institutional structure. Furthermore, it is immersed in a complicated network of psychological and socio-cultural interrelationships, an enormous complexity emerging from transcultural meetings and mutual influences. Their borderland-like character creates the potential space for transformation but it can equally become a source of troublesome dynamics that activates defenses. The question is then whether they lead simply to a change, something introduced into one system by another, or rather to transformation with all its implications as illustrated in the alchemical metaphor and with the emergence of a new creation involving both parties. Our reflections are based on a specific example from Polish culture that in anthropology is conceptualized as "the Great Borderland" between the West and the East. It not only amplifies the metaphor but also provides insight into its mechanisms.

The Reality and Dynamics of Router Training

Psychoanalytic Training

Any psychoanalytic training constitutes a particular situation. In some respects it is essentially different from any other in the field of helping professions. Usually it takes place within a space structured by the group, a professional association. It is shaped by basic assumptions about the human psyche which are meant to provide a holding space for individual and group dynamics. Upon completion, one becomes a member of a relatively closed and exclusive psychoanalytic society. Social institutionalization carries a shadow of elitism and power differentials. These may be in conflict with the basic hope of the candidates' capacity to transform. Many problems can then emerge, such as issues of flexibility in training and membership, the hierarchy within the society, training structures, etc. Discussions within the psychoanalytic world over the past few years have brought a questioning of different aspects of training. There is a sense of the need for revision.[5] Such questions and issues apply similarly to the Router Program. Because of the latter's unique institutional and transcultural character, the discussions will be different.

We believe that any Jungian training aims at deep transformation at a symbolic level akin to the process of initiation. It has the potential to serve this ideal or, on the contrary, to take on the aim of protecting the status quo, thus prohibiting transformation and not only individual transformation. From the perspective of the archetype of shamanic initiation, it is in the service of the community.[6] Trainees are inevitably influenced in the process. Influence brings change and if change leads to transformation, the goal can be achieved in a "good enough" way, however, in our view, the Router Program, because of its structure and character, creates particular transformational challenges for the trainee.

The Lack of Institutional Containment

Institutional training at an established institute has structures and sub-structures: the Training Committee, Reviewing Committee, Certifying Committee, etc. Usually, it has a spatial-temporal shape in the form of regular meetings and a physical dimension (a building in which seminars and meetings take place—a library, office, social room, kitchen, etc.). All are saturated with historical memories, generating a cultural collective milieu of tradition. This is missing in Router Programs, where the basic "institution" in training is the IAAP. This umbrella association does not provide all the essential components as there is no site. It has officers and committees who meet occasionally and then withdraw into their local communities. From the perspective of an individual router, they disappear as representatives of the institution. Router training usually takes longer than in an institute so such disappearances are frequent. Furthermore, individual analysts often refer to this community at large as *them* rather than *us*. We see how a local association remains the primary locus of identification. This already leaves individual routers feeling abandoned. The IAAP then is like a virtual organization, which can turn individual routers into virtual routers. This applies not only to individual routers, but also to post-routers (now IAAP Individual Members).

According to Kirsch, an analytic community carries several functions:

- a passage through the training with shared activities
- recognition by the analytic community as practitioners with adequate competencies
- a deep resonance about what has been learned about analytical psychology[7]

It is clear that for individual routers, the first two points are qualitatively different to those in an established institute, which perhaps creates a space for mutual fantasies affecting routers' levels of recognition. As individual routers we didn't go through these shared activities either. They are shared only in terms of their basic components: analysis, supervision, and theoretical training. When we looked closely and at a deeper level, we found meaningful differences.

Training Structure

A training structure is usually built on several core components: personal analysis, individual and group supervision, didactic seminars, group process, and evaluative processes. All are essentially different on the individual route training. In most cases, personal analysis is "shuttle analysis." Even if not, it is often done with a third person, an interpreter, in the room. Even if this is not the case, the language barrier still constitutes a challenge. The same applies to individual and group supervision. Both often happen at a distance on the telephone, via

Skype connection, or in a "shuttle" format. Didactic seminars usually involve translators, and access to basic literature can be limited.

The overall picture that emerges is a fragmented one. Under such circumstances, there is a deficit of coherence and consistency for the different elements of the training vessel. In such a situation, much of its function tends to be projected onto the personal analysis, which in turn makes for an added load of responsibility for the well-sealed training container/vessel.

A Field for Transformation

To the extent that the individual Router Program is essentially a transcultural experience, it is the bilateral meeting with "the other." It is concordant with the idea of initiation, which is a journey into the "realm of otherness." It is a structural setting for meetings at the frontiers. This kind of psychological space has a certain liminal quality, typical for an experience in the borderland, without which no permeability is possible. There is a deep exchange at the level of the cultural unconscious. We know that psychic complexes can present either a threat or an opportunity and perhaps our own cultural complex of the Great Borderland makes us particularly vulnerable in this kind of training; however, it also helps us recognize the dynamics behind the processes embedded in the Router Program.

Initiatory processes need solid containment. Rites of initiation are usually provided by the society of elders contained by cultural tradition. In the case of the Router Program, both "the elders" and cultural container are virtual and outside the cultural context of the initiates. There is no doubt that individual routers are trained and eventually become Jungian analysts. It seems to us that the Router Program demands an explanatory metaphor in which both parties are in the process of transformation. Analytical psychology naturally provides such a metaphor with the alchemical processes and the idea of an interactive field.

In his paper about teaching candidates in the North Pacific Institute, Ladson Hinton described the "immersion experience" in the training oriented towards the process of transformation.[8] Candidates, besides memorizing "the known" are encouraged to think critically and creatively. This is sometimes painful and anxiety-provoking. He wrote,

"One could see the goal of the education, similar to the goal of analysis, as encouraging the unveiling of a potential space, a 'thirdness.' This involves tolerating uncertainty. ... Without this potential space one is stuck in a world that is fixed and concrete."

The concept of an interactive field suggests that in the analytic process there is an emergence of the "third," a field holding the analytic pair, not reducible to its parts. This is an emergent structure, autonomous and purposeful. If we take Hinton's thinking further, what is typical of the analytic relationship is also present in the training situation. It helps to understand the reciprocal quality of the training processes, which avoids the trap of reducing the field phenomena to failures and individual and collective pathologies, mistaking *borderland* for *borderline*.

From studies of the interactive field we know that it is purposeful and aims at the creation of new structures that could be called *Self-structures*.[9] These new structures emerge and are unpredictable. In the training situation, they are confronted by the conscious and unconscious expectations of the trainers, who may, for example, have expectations to form a new institute and training facility from the Router Program. It has happened that what was emerging in Router Groups did not meet the expectations of the trainers. The question is not only what the "elders" bring but what they expect to gain from their investment in training. If we take the idea of mutual influence seriously, we must also acknowledge that destruction of old forms precipitates an emergence of new ones. This can be a tumultuous and chaotic process involving both parties.

We believe that careful analysis of the nature of new training structures can help us to avoid thinking in terms of "successes" or "failures." We have heard many rationalizations such as the router training is not good-enough, post-communist countries are incapable of organizing a democratic group life, they carry cultural wounds too deep to engage "properly" in training, and so on. Only when we try to suspend certain kinds of judgments can we then begin to see the complex network of meetings, interactions, confrontations, and influences that make up a Router Program.

Personal Transformations

For both of us, our training was a deeply transformative experience. At the same time, we became painfully aware of the demands and consequences of this transformation. It raised questions about the factors that make transformation possible and those that can lead to a situation of "failed transformation" or into a position of being influenced and changed at the level of persona rather than that of self and culture. Transformations are meaningful for both parties and also for the culture as a whole including the field of analytical psychology. We would like to look in more detail at some aspects of transcultural meetings. They lead either to transformation or to the inhibition of this process leading to feelings of abandonment and experiences of a failed dependency understood by group theorists as a traumatizing experience at the group level.[10]

Paraphrasing Heyer, we can say that cultural underpinnings always influence our thinking about training—both the way we are being trained and the way we train others.[11] They influence our perception of the "other," the way we encounter otherness and build it into our own identity. These processes are framed by more or less shared experiences of the trainer and the trainee, and their meeting at the frontier creates a completely different experience where cultural characteristics are treated relatively. It is a place where the typical ego-oriented Western consciousness that gave birth to psychoanalysis does not easily embrace issues of language, myth, symbolism, history, and cultural difference.[12] Such situations require a completely new attitude of not-knowing on both parts. Otherwise, it is likely that one point of view will intrude into the other. When this happens, regardless of how, genuine meaning is lost *in statu nascendi*. As Horne writes, "From the moment we take our first breath, we are inevitably and unwittingly immersed in the context of our micro-culture, about the nature of the world, and the correct practices with which to engage it."[13]

The learning process requires deconstruction of already acquired knowledge about the world. Here, the situation becomes more complicated. As an example, we refer to the teaching on a Jungian training in Poland by an English analyst. This can be seen as a meeting of "otherness" in the following ways:

For the teachers:

- "going to the East" with all its complicated historical, cultural, and symbolic implications, fantasies and assumptions
- complicated language issues (taking part in a translated seminar, the sound of a foreign language, uncertainty about the accuracy of translation, etc.)
- physical cultural space (the trip to the seminar, the rhythm and customs of the foreign country, architecture, etc.)
- experience of the way public and personal space is mediated (different customs about greetings, physical distance, emotional expression, etc.)
- visiting analysts' own fantasies of the foreign country, historical knowledge about its past, knowledge and images about its present based on media information (for example an image of Poland as traumatized)
- images based on meetings with Polish people working in the UK
- attitudes to what is foreign, part of one's own cultural complex and unconscious projections
- need to "save" or heal a traumatized country, assumptions about trainees' wounds and traumas (activation of the archetypal wounded healer dynamic without a possibility to really work with it within the given frame for the training)

For the trainees:

- stepping out of the mainstream of psychotherapy training into an area of knowledge that generates fantasies based on fragments of Jung's translations (inevitably projections of something that is being searched for without clear recognition of what it might be); stronger than usual experiences of "not knowing" at the beginning of the training
- meeting teachers that come from "the West" with their complicated and different historical,

cultural, and symbolic heritage including their fantasies and assumptions
- projection of anticipating something "better" than can be acquired from teachers "from here"; the activation of shame lying in the cultural unconscious with its roots in the history of twentieth century communism
- experience of different social and cultural attitudes and behavior and the way personal space is shaped by the teacher
- language issues including the need to communicate and having to rely on the translator as the mediator, and the inevitable frustrations this will bring
- for English-speaking trainees, the experience of talking about themselves and building psychological knowledge using different grammatical structures and semantics
- images of visiting analyst's country of origin and unconscious projections of cultural complexes and what is foreign
- activation within the learning process of unconscious areas of transgenerational and cultural trauma, activating old wounds and defenses

These represent the complicated network of conscious and unconscious interactions that will inevitably take place. Using the alchemical metaphor, if transformation is to happen, then both parties need to "undress" from their own cultural experiences. This process can be demanding for both, but, with some continuity, the transformational space may be maintained, however, teachers usually alternate with their colleagues coming also from different cultural backgrounds themselves and who may not be there for long enough to sustain a strong frame that would permit a full understanding of a transcultural meeting. The trainee can be left with an experience of fragmentation from such different cultural encounters. This means then that personal analysis is likely to provide the most secure container but of course, it is also taking place in transcultural space, usually in a foreign language, as "shuttle" analysis, etc.

The "router coordinator" appointed by the IAAP to help the group could be a good container but this function is most usually set up to provide organizational help. Moreover, it is offered "from outside"; the coordinator liaises between the IAAP and the trainees. The borderland experience evoked in meetings with the cultural unconscious and cultural complexes is then unlikely to be equal for each side. While the teacher or coordinator is "visiting the borderland," the trainee "lives in the borderland." This hypothesis is particularly poignant in the context of Polish history and experience because of its location between the West and the East.

Areas of meeting connect and disconnect depending on the respective cultural spaces of the trainee, teacher or analyst, and other factors described above. How will this affect individual and group development within the training?

THE PARADOXICAL CHARACTER OF A MEETING IN THE BORDERLAND

The process of training in analytical psychology involves development in the use of the self in clinical practice, coming from the integration of experiences of personal analysis, theoretical study, clinical experience, supervision, self-assessment, and learning within the community.[14]

Successful progress is likely to lead the trainee deep into the primitive areas of his or her psyche. The process of transcultural training leads down into the cultural unconscious as well. What is often found there are the traumatic, psychotic-like residual experiences and images from the collective history and the primordial elements of endogenous heritage which may have been lost. The more they have been lost, the less visible they will be. They remain a secret psychic attitude "deeply buried, experienced by all and observed by none."[15] They are unrecognized and can be acknowledged via their absence rather than their presence. When acknowledged, the chances are there will be a recognition of these haunting experiences by "the society of the dead," standing behind our backs and asking for resurrection to participate in mourning past and current losses embedded in collective development.[16] In the case of Poland, we are confronted with the task of regaining our lost multiculturalism, the hidden wealth and tapestry of our borderland, once eradicated by history and national tragedies, by wars and genocides, and finally by the poverty endured during the communist regime.

What is less visible exerts more influence unconsciously and, if not adequately recognized, can be played out in the dynamics and fate of the group. Mediating the borderland position involves not just bridging different languages but also finding a new and emergent language. This new language introduces new ways of communicating and involves finding grammatical boundaries and semantic differentials that permit meaningful communication (the content).[17] Communication about personal and collective experience usually takes place within the language that is specific for the culture and recognized intuitively by both sides. In the case of a transcultural experience, this process will take a different form.

For one of us, the moments in analysis when the English language was incapable of containing the Polish fragment of experience were recognizable by a breakdown of syntax. While speaking in English seemed to be fluent, the syntax started to take on more of a Polish form, making the meaning less and less comprehensible. It was then amplified in a repeated dream about being in the role of interpreter and guide for an older man visiting Poland, who looked much like Jung and was a Jungian analyst. When moments of difficulty or failures to embrace foreign cultural experience come up, they can be recognized from their content and in the structure of what is happening (the syntax). We might say that the new language needed to embrace this situation is neither Polish nor English. It is something that has emerged in the area between—in the borderland.

In the example above, this new space was filled with possibilities to work on traumatic experience emerging from the cultural unconscious and resulted in creative writing from the dreamer. The dream highlights the essential factor that contributes to this area of transition: a fragment of shared experience and familiarization with "otherness" that results from transcultural meetings and serves an intermediary role for both parties.

The word *interpreter* derives from Latin *interpres* which denotes an intermediary presence of a negotiator, mediator, or messenger. We can see in the above example how embracing a fragment of one's own psyche facilitated a *rite de passage*, and a return of a domestic cultural experience that had been previously out of reach—an aspect of traumatic heritage locked in the cultural unconscious.

One of us who trained in the United States returned to Poland and dreamed of standing on the street between two aboriginal figures: a male Native American and the female Polish Highlander, bringing them both into his childhood home. This is a clear image of the borderland position. Here too, the transition was viable because of the availability of an intermediary space; the dream setting was located on a pedestrian crossing with traffic lights making the passage possible and safe.

Both these examples show the effects of personal work and the creative outcomes of interpersonal exchanges in a transcultural setting. To come back to the metaphor in our chapter title, such meetings and subsequent reflections about the mutual changes that took place meant that there was potential psychic transformation, but this process can take other turns when the negative aspects of complexes are activated. The master of the borderland is Hermes, the duplex and treacherous deity. When negative complexes are constellated and met with deficits in the interactive field, there is insufficient containment to lead to transformation.

One example would be the clash between the mother complex with cultural complexes from English and Polish culture that took place early in the Router Program. A liaison person once commented that the group held an unconscious image of her as Lady Pamela arriving in her pink Cadillac and bringing deliverance from difficulties (a mother complex figure). This image felt foreign and incomprehensible to Polish trainees. In response, it evoked complicated feelings toward the liaison person as a superior figure. It constellated the Polish mother complex who is out of reach and brought to the surface unconscious guilt for making the mother suffer. This experience can be seen as anchored in the archetypal fantasy of Poland as "the mother," dominated by an image of *Mater Dolorosa*, Our Lady of Sorrows. It was further complicated by following projections from the British that Poles cannot organize themselves democratically and from the Poles who thought the British were acting out their imperialistic needs. Two powerful motifs from national cultural complexes were activated: the abandonment of Poland in the face of threat and the loss of the British Empire. The duplex (Hermes-like) character of an image of the Lady

Pamela is worth noting here. It refers to an "extra-terrestrial" (the ultimate "otherness") quality of the pop-cultural figure of Lucia Pamela, "the Moon Lady." But it may refer to the Lady Pamela Hicks, who epitomizes the British aristocracy and is thought to be the last witness to Indian Independence Day in 1948.

A series of such experiences were eventually followed by further difficulties and, ultimately, by a split within the group. In its early development, the management of the Router Program and the life of this particular group seemed to contribute to unmediated individual and collective cultural experiences, which stirred up complex organizational dynamics and left the group saturated with undigested traumatic experiences—traumatization at the group level may be understood as a failed dependency.[18]

In Poland, such re-constellations again fall into the familiar territory of the national complex of the Suffering Hero. It carries a Messianic and Promethean character expressed in the famous slogan of Romanticism: "Poland as a Christ of The Nations."[19] This cultural complex emerged during the time of Poland's partitioning in the nineteenth century and remained alive in the subsequent struggles for independence. Although at times it carried defensive associations, for the most part, it saved the traumatized soul of the nation. It might be seen then as a source of Polish resilience. Essential in this aspect of Christ's mythology is "the harrowing of hell," the moment *between* death and resurrection. This, however, requires the stepping down from the Cross and immersion in the experience. The Christ complex operates by preventing transformation. Transformation requires the kind of authentic suffering embedded in the death of an old structure which seems to be part of the experience in the borderland and demands an act of relinquishment. Abandonment by visiting teachers distressed by the energy of the Polish cultural complex exposes the group to re-experiencing trauma at the mythical level; "Father, why hast thou forsaken me?" Then things get out of proportion, unmediated by a sufficiently strong structure of collective consciousness. The group avoids independence (experienced at an unconscious level as death) by remaining dependent (neither dead nor alive). The process of transformation is prevented, leaving the group alone and abandoned with its *eternal return of loss*. As teachers for the groups, we witnessed the reactivation of those old patterns in the dreams brought into social dreaming matrices, when the group psyche was searching for new solutions.

To an outsider, the process illustrated in the examples above may seem tumultuous and even destructive. This kind of confrontation between two perspectives—internal and external—challenges the forming group with fantasies of how it could or should be. For example, the group is supposed to develop into a professional institute and, eventually, into an autonomous training organization; the group is not supposed to split and so on. Images of old forms are projected into new structures that are still in the process of emergence and where it is not possible to predict a final form.

We need to be able to hold the meaning of the borderland in Polish culture as prevailing in all transcultural meetings, acknowledging cultural complexes on both sides. Then a new space for reflection may emerge, facilitating integration and new possibilities. New structures emerging in Router Groups may not turn out as anticipated but not because they have failed. Their character may not resemble anything already known. It is not just a matter of a new structure replacing an old one; rather, there may not be a structure at all—in the usual sense of the word. We are living in times of fluid identity with the focus of attention shifting from structures toward parts and relationships between parts.[20] We may not yet know the working models for the new kind of reality in which we live. Ultimately, Router Programs may be viewed as virtual training, emerging organically within the IAAP— also a virtual organization in times of global virtualization. Given the IAAP's *modus operandi*, can we expect it to produce structures similar to those that develop in the local institutions all over the world? The question remains as to whether we can expect Router Programs to have a life analogous to the lives of institutional trainings.

Transcultural experiences are no longer only the province of travelers. Thanks to modern technology, we are all travelers, irrespective of our wishes and intentions. The world is changing at a rapid pace so that we participate in transcultural meetings every day. We have tried to bring to life the complex processes and dynamics to be found in router trainings and to emphasize that they may tell us something more than just about those cities and countries where psychoanalysis was previously banned. They tell us something about the need to approach transcultural space in a different way. This demands connection from all involved with transcultural experience, and to find the capacity to "undress" so that their potentially transformative natures may be known by both teachers and trainees.

Notes

1. James Astor, "Supervision, Training, and the Institution as an Internal Pressure," *Journal of Analytical Psychology* 36 (2, 1991): 177–91.

2. P. Homans, "Foreword," in Thomas Kirsch, *The Jungians* (London and Philadelphia: Routledge, 2000), p. xii.

3. Thomas Kirsch, "IAAP and Jungian Identity: A President's Reflections," *Journal of Analytical Psychology* 40 (2, 1995): 235–48; and Kirsch, *The Jungians*, pp. 165–220.

4. Tomasz J. Jasiński and Malgorzata Kalinowska, *Differences of Fantasy, Fantasy of Differences: Reflecting on Jungian Identity in the Context of the Individual Route* (paper, *Journal of Analytical Psychology*, XIIth International Conference, Berlin, 2014).

5. Astor, "Supervision, Training, and the Institution as an Internal Pressure," pp. 177–191; Patrick Casement, "The Emperor's Clothes: Some Serious Problems in Psychoanalytic Training," *International Journal of Psychoanalysis* 86 (4, 2005): 1143–60; Angela Connolly, "Some Brief Considerations on the Relationship between Theory and Practice," *Journal of Analytical Psychology* 53 (4, 2008): 481–99; Catherine Crowther et al., "Fifteen Minute Stories about Training in Russia," *Journal of Analytical Psychology* 56 (5, 2011): 627–52; César Garza-Guerrero, "Reorganisational and Educational Demands of Psychoanalytic Training Today: Our Long and Marasmic Night of One Century," *International Journal of Psychoanalysis* 85 (1, 2004): 3–26; Ladson Hinton, "Teaching 'Origins of Depth Psychology': Didactic Overview and Candidate Experience," (paper, *Journal of Analytical Psychology*, VIIth International Conference, Montreal, 2006); Otto F. Kernberg, "Thirty Ways to Destroy the Creativity of Psychoanalytic Candidates," *International Journal of Psychoanalysis* 77 (5, 1996): 1031–40; Michael Horne, "There is No 'Truth' Outside a Context: Implications for the Teaching of Analytical Psychology in the 21st Century," *Journal of Analytical Psychology* 52 (2,

2007): 127–42; Tom Kelly, "The Making of an Analyst: from 'Ideal' to 'Good-Enough,'" *Journal of Analytical Psychology* 52 (2, 2007): 157–69; Jean Kirsch and Suzy Spradlin, "Group Process in Jungian Analytic Training and Institute Life," *Journal of Analytical Psychology* 51 (3, 2006): 357–80; and Joann Ponder, "Elitism in Psychoanalysis in the USA: Narcissistic Defense Against Cumulative Traumas of Prejudice and Exclusion," *International Journal of Applied Psychoanalytic Studies* 4 (1, 2007): 15–30.

6. Joseph L. Henderson, *Thresholds of Initiation* (Wilmette, IL: Chiron Publications, 2005); and John Merchant, *Shamans and Analysts: New Insights on the Wounded Healer*, (London and New York: Routledge, 2012), Kindle edition, chap. 1.

7. Kirsch, "IAAP and Jungian Identity," p. 247.

8. Hinton, "Teaching 'Origins of Depth Psychology,'" pp. 91–92.

9. N. Schwartz-Salant, "On the Interactive Field as the Analytic Object," in M. Stein, ed., *The Interactive Field in Analysis*, vol. 1 (Wilmette, IL: Chiron Publications, 1995), pp. 1–36.

10. Earl Hopper, ed., *Trauma and Organizations* (London: Karnac Books, 2012), Kindle edition, Foreword.

11. Gretchen Heyer, "Caught between Cultures: Cultural Norms in Jungian Psychodynamic Process," *Journal of Analytical Psychology* 57 (5, 2012): 629–44.

12. Jerome Bernstein, *Living in the Borderland: The Evolution of Consciousness and the Challenge of Healing Trauma* (London & New York: Routledge, 2006), pp. xv–xix.

13. Horne, "There is No 'Truth,'" p. 130.

14. Jan Wiener and Christopher Perry, "Fifty Years On: A Training for the 21st Century, from Couple to Community," *Journal of Analytical Psychology* 51 (2, 2006): 227–49.

15. C. G. Jung, "The Meaning of Psychology for Modern Man" (1934), in *The Collected Works of C. G. Jung*, vol. 10, eds. and trans. Sir Herbert Read, Michael Fordham, Gerhard Adler and R. F. C. Hull (London: Routledge and Kegan Paul, 1964), § 315.

16. Samuel Kimbles, *Phantom Narratives: The Unseen Contributions of Culture to Psyche* (London: Rowman & Littlefield, 2014); James Hollis, "The Ghosts of Our Parents," in *Hauntings: Dispelling the Ghosts Who Run Our Lives* (Asheville, NC: Chiron Publications), Kindle edition; and Greg Mogenson, "The Resurrection of the Dead: A

Jungian Approach to the Mourning Process," *Journal of Analytical Psychology* 35 (3, 1990): 317–33.

17. Dale Mathers, *An Introduction to Meaning and Purpose in Analytical Psychology* (London: Routledge, 2001), p. 161.

18. Hopper, *Trauma and Organizations.*

19. Malgorzata Kalinowska, "Suffering Hero—Messianism and Prometeism in Polish Cultural Complex," in eds. T. Singer and J. Rasche, *European Cultural Complexes* (New Orleans, LA: Spring Journal Books, in press).

20. Zygmunt Bauman, *Identity: Conversations with Benedetto Vecchi* (Cambridge: Polity Press, 2004).

CHAPTER 10

Bridging Two Realities
A Foreign Language in an Analytic Space

Kamala Melik-Akhnazarova

Voyage Voyage

At the early stage of my mid-life crisis, when I was trying to change my profession into a more satisfying one, I was "by pure chance" invited to interpret the initial interview for the candidates of the International Association for Analytical Psychology (IAAP) Moscow program for analytical psychology. I began that job as a total stranger, as an observer, but very soon found that I was plunged into the process. Eventually, I experienced all the stages of that program in concentrated form. It helped me not only to reconsider my experience of being an interpreter and thus to bring it to a close, but also (thanks to a stable positive transference) to obtain a new professional identity by training as an analyst. I came out of the shadows and dared to find my own place in the family of analytical psychologists; so now I am practising the art of interpretation in its other dimension, in the next turn of the spiral of life.

The task of writing a chapter about my experience caused a lot of inner questions—first of all, shall I write this piece initially in English or in Russian and then translate? Which is better for me? What is easier and how does it affect the sense? What are the resulting changes? All these questions that crowd my head are imbued with meaning and invite one to a linguistic and semantic journey. Learning a foreign language gives one an opportunity to travel far away, not only in one's body but in one's mind as well. (It is interesting that in the former Soviet Union, the country that closed its citizens off from any exchange with foreigners, the academic school of teaching and learning languages was very strong—and to my mind it was a way for people to "travel in the mind" and to hold an "inner opposition" to the totalitarian state.) And any journey is a chance to see the world through different eyes.

Applying this to a foreign language, we may say—to hear the world differently or to voice it differently, to perceive it differently, to become acquainted with another facet of it and maybe even to let oneself become "another" for a moment, to change, to become open to the agent of change within oneself—an impulse for transformation is given.

Bridge Over Troubled Water

It is curious that the Russian Revival Project that was so active in organizing the IAAP training program in Russia used an image of the St. Petersburg split-bridge as its symbol. At that time, at the end of the 1990s and early 2000s, the two arms of that bridge were coming together and everyone involved felt excitement and pride and looked forward to witnessing the dawn of a new Jungian reality. Now the political situation is such that the arms of the Russian-Western bridge are separating again—no one knows for how long. But the new body under the name of the Russian Society for Analytical Psychology exists, so the waters of the unconscious between the two arms of a symbolic bridge carry within them a precious baby.

The Russian Revival Project was an ambitious and risky plan—and maybe the high level of anxiety was part of the shadow of the whole partnership deal. On the one hand, there was inspiration, the wish to share wisdom, to elucidate, and on the other, a desire to amplify knowledge, admiration, striving to learn and to improve the existing situation. These feelings had their unconscious dark side. Projections showed a tinge of superiority-inferiority complex—the missionary's infallibility and covert remonstrance of the "flock," mutual fear, and watchfulness in the face of all possible inequalities (not to mention the vestiges of the previous Cold War period and the differences between so called "democratic" and "totalitarian" mentalities). Russia is a sample of instability, revolt, and split identity, characterized by brilliance and extreme destructiveness at the same time. Britain is the world symbol of tradition and conservatism that borders on rigidity and resists the emergence of anything new. In the clash of these two clichés, what would be born? How would they cooperate, both inwardly in the intrapsychic realm and outwardly on the social plane?

I guess no one knew exactly, however, I became an interpreter for the supervision sessions of the project, and the role of interpreter *volens nolens* is a personification of Hermes, the guard of the boundaries and divine messenger. Mercury, that quicksilver of alchemists, was working hard, i.e., the archetypal level of the encounter was, to my mind, activated more often than during the usual direct conversations—be they analytic sessions or supervision. The whole situation was just more intense, and it was reflected on many different levels.

First, the obvious: the triangulation of the dyad analyst-analysand (or supervisor-supervisee). This aspect was referred to by Elizabeth Gordon when she described her own experience of being an analyst via interpreter in St. Petersburg.[1] It was no longer a one-to-one dialogue, but the conversation of three persons, yet with different roles and positions. Is it possible to find some universal meaning here or should we interpret such situations differently each time depending on the concrete context? The key question is how does this triangulation affect the interpretation of an analysand's personal story? Jung's familiar diagram of transference-countertransference projections and cross-projections becomes, in this case, more complicated. If we try to make a graphic figure of it we can draw an octahedron with multiple directions of interchanges. Here the conscious and unconscious communications of an analyst and analysand (or supervisor-supervisee) are supplemented with conscious and unconscious communications of those two with the interpreter and all three between themselves.

In this more complex and three-dimensional construct the interchanges are intensified and the directions of

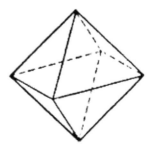

Figure 1. Three-dimensional adaptation of Jung's diagram of transference-countertransference projections and cross-projections.

those interchanges become more various and chaotic. The peculiarities of real practice, of course, depend on actual personalities and conditions; the existing transferences and countertransferences could be positive or negative, but, in any case, this model is useful for reflection and for taking into account those aspects of the work that otherwise would escape awareness.

Lost in Translation?

The first thing that I reflected upon was how the presence of the "third" in between two speakers altered the time picture and transformed the prosody of speech. Reactions are no longer spontaneous and immediate but are subject to pauses and repetitions (which threaten enormously the presence of humor). The content is run through at least twice, being pronounced in two different languages. The speaker, no matter who is addressing the listener, does not make the meaning, i.e., the sense-making process is split, the meaning voiced by the other. The parties involved in communication are forced to exhibit patience, tolerance, and trust, which is not always easy because the emotional load is high. I can recall those moments of significant anxiety during the training program, such as the key moments of evaluation at the different stages (interviews, annual reports, exams, etc.). At such a time an interpreter played the role of a buffer or screen for the projections of fear, envy, and anger and became a target for accusations of faults, mistakes, misunderstandings, and misinterpreting; while in those calm and clear moments of mutual respect and admiration, an interpreter received all kinds of positive, idealized projections and flattering compliments. This shows that the transference picture is split but at the same time belongs to the same container, i.e., the split part is acted in and not acted out but still avoids attention.

The interpreter, from her part, has to be alert, quick, and technical. The conscious concentration of all three to be explicit, meaningful, and eloquent intensifies the process and flurries the unconscious. This mechanism works as a vacuum pump or a psychopomp for the meaning to be brought in or to emerge (which also refers to the archetype of Hermes). Here we witness an interesting phenomenon, namely, the exclusion of the interpreter. Not only is he the one who bridges the sense-making process, but at the same time his personality is being notably excluded from the process. Even in the analytic field, with its attentive focus on the minute details of what happens between analysand (supervisee) and analyst (supervisor), the figure of an interpreter avoids such attention—there is no time and space either for her conscious responses or for processing unconscious reactions. And here the tricky and thievish aspect of the archetype of Hermes takes revenge. An interpreter may simply lie—unconsciously, most of the time. But there are moments when it is tempting to slightly alter the nuances of meaning—by the choice of a word, an involuntary comment, an increased or biased emphasis, overt emotionality, or unnecessary irony. The original speaker then becomes an innocent victim with no possibility to remedy the damage.

At this point we come closer to another archetypal image that is powerfully constellated in any work via an interpreter, an image that refers to narcissism and is split between grandiosity and void. In the myth of Narcissus, the nymph Echo was transformed by jealous Hera into a creature with no voice. Later she could only appear by repeating (or echoing) someone else's phrases. Echo, enamored with Narcissus who paid no attention to her feelings, melted away because of her love for him, unable to express herself. In 2008 Nathan Schwartz-Salant, lecturing in Moscow, made an eloquent analysis of this part of the myth.[2] While interpreting this section of his speech I suddenly sensed very vividly how this image was linked to the archetypal core of the profession of the interpreter, to the unconscious drive that gives propulsion to such professional fulfillment; it was a kind of embodied insight.

Thus we have an indication of some deep feeling that cannot find verbal expression—either not heard or not understood—and the whole deal revolves around this issue, weaving the fabric of an encounter. Even more, there is also an absence (something that is not yet embodied) that carries a significant aspect of the meaning for the intersubjective field.

An interpreter who acts as mediator and agent often feels an enormous craving for attention and is under pressure to express something inexpressible. While the conscious efforts are rewarded by the actual process of communication through his skills and the accompanying feeling of satisfaction and of being needed, there still remains an unconscious ingredient—the suppressed negative feelings around the issue of being used only as a function, bringing a great deal of envy and competitiveness. This ingredient often remains unacknowledged. Being mostly unconscious, how does it affect the field of an analysis or supervision?

My own experience of interpretation belongs only to supervision sessions. While supervision provides a safe container for an analysand by adding another level, another containing envelope, to the process, here the complex issue of "absence" is even more intensified. The patient, the main protagonist of the discussion, is physically absent but remains the main focus of attention of the dyad supervisor-analyst, while the interpreter, the carrier of the language and the sense-maker, is physically present but emotionally absent. So, can we say that this issue of absence can be something meaningful for the material of an analysand just by the fact that it is so brightly constellated in the parallel process of supervision? Maybe not, but at least we can pay attention to this absence and try to reflect on it in all its possible contexts.

When the meaning is not easy to find and is reinforced by the immediate situation of working via someone else, the unconscious picture is charged with a great deal of irritation and anger and the figure of the Other is exaggerated. It reminds one of a sandwich, where two pieces of bread cannot meet each other without the third layer in between (this image as an illustration for the unconscious dynamic of a parental couple was used by Kenneth Reich in his Moscow lectures).[3] Doesn't it look like a parental couple (maybe Hera and Zeus with their constant battles) unable to build a trustworthy relationship directly and thus unwittingly using a child (an "innocent" outsider) to communicate their love and hate to each other? In such couples the child is often viewed only as a function and experiences a bitter lack of attention from the parents. The child's conscious attempts to become good and noticeable are counterbalanced by unconscious anger and destructiveness.

Feelings ... Nothing More than Feelings

One of the first actual contentious issues for me as an interpreter in the program was of my emotional involvement in the content of the discussed material. The task of the interpreter is to maintain emotional involvement to just such a degree as to be able to provide a correct and prompt exchange of information. She should understand emotionally-charged material correctly at the same time as exhibiting a capacity to maintain an emotionally-distant position in order not to let her own feelings interfere with the process. The suppressed ego of an interpreter strives for recognition. Wounded pride is compensated for in different ways. Frequently, it is the reason for over-dramatizing the content, for giving emotional accents that do not exist in the original version, for using striking expressions, or if one's personal temperament is different, using pauses, comments, and inexact, generalized interpretations. Sometimes the interpreter sinks into a trance-like dissociation. All these reactions, no doubt, confuse the process. But at the same time they become an integral part of it, and if acknowledged, could promote meaning. The most common conscious and intellectual attempt on the part of the interpreter would be to try to avoid being overwhelmed, to keep transparency and neutrality. But this task is not fully feasible. Instead, if we understand the whole *mise-en-scène* as a situation of "total" transference and pay more attention to the factors that may cause confusion in interpretations, we find the lost and sometimes very important ingredients of the meaning.

This is illustrated in the following emotional impressions of personal analysis via an interpreter:

> The interpreter was involved in the process emotionally and I had a feeling that I was observing myself from aside, as if I was trying to understand what were the feelings of this person who reported 'my meaning.' The feedback from the analyst also fell in two lots—to me and to the interpreter. I had a wish to 'become friends' with the interpreter—she became a transitional object. The separation from my analyst was unbearable: my infantile trauma was reproduced. The interpreter became a lively embodiment of a psychological obstacle in the relationship with my mother. An absence of understanding with my real mother was symbolically reflected in the unknown language of analysis. But at times (probably due to the analytic space) I had a feeling that I was understood without those foreign words that we used to talk to each other. (Elena Rezanova, personal communication, September 2014).

Another interesting aspect of the work of an interpreter in a supervision setting is the inevitable personal interaction both with a group and with the supervisors. However small those could be, they cover a strongly felt gap and give an interpreter a sense of belonging. Even with our best intentions not to have social relations for reasons of confidentiality, there is a strong wish to become closer. Besides having the role of a guide to the cultural space of each country (and here again the archetype of Hermes/Mercury is activated), there is also an interesting aspect of "betrayal." Jacqueline Gerson addresses this idea: "the common, talked-about Mexican phenomenon of *Malinchismo*, which refers, in my country, to disliking one's own and preferring the other—giving oneself to the foreigner and abandoning and betraying one's own."[4] Gerson refers this tendency to a Mexican cultural complex, but it is very much the same for Russia as well, with a half-conscious tendency to negate one's own and to extol the other, so that the interpreter was inflated as if by closeness to "gods."

Another facet of the interpreter's role was of a stabilizing figure, who worked to maintain the safety of the group; he was the constant object while the supervisors replaced each other. (The monthly supervision process was organized so that each supervisor appeared on a three month rotation, while the interpreter was the same every month.) For the supervisors, who would share their few moments of leisure in the breaks with the interpreter, she became "no stranger" in the foreign country and thus a stable and secure object. These personal relationships that were continued beyond the frame of the supervisory process, on the one hand, watered down the boundaries and, on the other, warmed up the formal process and helped to take the sting out of the interpreter's narcissistic wounds.

So, my first hypothesis is that the interpreter represents two different archetypal images that are linked to the process of such work, both of them extremely important. First, there is Hermes—the guardian of boundaries and facilitator of an exchange, the psychopomp. Second, there is a powerful dynamic of narcissism—the evasive object of love provokes the other's violent urge to break through the void to a psychic existence in the eyes of the beloved creature. It also can be viewed as an evocation of the preverbal stage of development. Unless we are aware of this powerful unconscious dynamic we are subject to numerous mistakes.

Analyzing This and Analyzing That

If an interpreter becomes a screen that absorbs the essence of the intersubjective field and to some extent enacts the characteristics of this field, becoming in a certain degree an unconscious personification of the analytic third, then what happens if analysis is held without an interpreter, but in a language that is not a mother-tongue either for analysand or for analyst? In a modern globalised world when migration is a common fact of everyday reality this is quite a possible situation. I had an opportunity to experience both sides—in my own individual analysis and in my work as an analyst with foreign analysands in Moscow—and in both cases English was the analytic language.

People Are Strange ... When You're a Stranger

The flow of my analysis happened to be barred by rapids and I overcame it through a change of analyst—I changed the Russian "mother" to the foreign "father." In my case the choice was influenced to a great degree by processes that were taking place in a bigger, collective entity—the Jungian family of the Russian Society for Analytical Psychology (RSAP) at the start of its existence. Those processes were inevitably interfering with the space of my individual analysis. It felt similar to the changes that happen in a family container—the crises of development that affect the individual life of each member of the family as described in the conception of *anamorphose* by Pierre Benghozi.[5] In my personal experience of this episode, I was unconsciously reproducing a family situation when at the age of five (due to powerful family perturbations) I was sent away from my parental home to live for a few months with my grandmother. Those perturbations in my family set up a certain constellation in my psyche—an alienation from everything maternal.

Linguistically, it was a peculiar situation—the new analysis was held in a southern European country (not English-speaking), but in English, because my knowledge of the local language was on the pidgin level, not sufficient (I supposed) for analysis. The English of my analyst was good but his pronunciation was sometimes unusual for my ear, so in the beginning those moments when I did not understand some word or phrase were quite frequent, and I would ask—or not ask—him to repeat. He was not speaking his mother-tongue with me either and I suppose that in a certain degree his habitual elegance of speech was limited. So the language became a bright signal flag: "Keep vigil! Here is the boundary, the total stranger. He speaks differently, i.e., he thinks differently, feels differently, perceives differently." At the same time I was plunged into a cultural and linguistic milieu that was very nurturing emotionally and inspiring but did not understand any English (not to mention Russian). It was a small town, practically a village, and the surrounding nature became for me a perfect archetypal container, combining the features of my Moscow dacha (the nest of my childhood) and the mountains and forests of Caucasia (the motherland of my ancestors). In such a place I immediately found myself in the role and age of a child acquiring the first skills of speech.

For a long time I took delight in such a state and had a special pleasure in sharing the emotional communication outside any semantic involvement, socializing non-verbally and giving myself a break on the conscious level. In some ways it was the extreme opposite of the work of an interpreter, whereby conscious verbal skills are intensified and the possibility of a reverie-state is reduced. This situation reproduced my past—those times when relatives from Azerbaijan came to visit us in Moscow. Then Azeri speech would flood our home and I did not understand but was willingly and joyfully bathing in it. The family became numerous, full, diverse, multi-branched, fully constructed.

Every morning I made my everyday way to visit my analyst—just as I used to go to school in my childhood (to my English school that directed my socialization up to the midlife crisis when I no longer wanted to correspond to the role my father "prescribed" for me, and I swung over to psychology).

And still I was seeking for the "father" and searching his aid in the process of a second quest for socialization, and I was talking English. My analytic father also seemed to me too strict from time to time and distant; he raised the bar too high, his interpretations, profound and exact, outsprinted my understanding. I knew that it was good, that it provided me space for growth. Was it not what I was striving towards? But I liked him most at those times when he was silently slumbering in my sessions; when I just broke off in silence, lending an ear to his breathing like a child comforted on the lap of a parent, then I knew that no matter whether I was talking or not, what I was talking about, or in what language I was talking, I'd be understood, that I was in firm hands and that we were together.

Eventually I learned his language well enough to participate in everyday communications and to build very warm, vivid, and reliable relations with people around me. But I never dared to start talking his language in my analysis. Why so? Somehow I did not want it. "My" English became "our" English; it was a language that differentiated analysis from the sonorous accord of the surrounding sensual perceptions. It helped me not to sink into (or not to be devoured by) a steamy, vibrant, ever-changing sea of tastes, smells, visual impressions, and sounds—limitless, captivating, seductive. It helped me to build an ark to keep me alive and safe within that flood of sense impressions, to be protected and to have something to lean on; besides, it was "our" language in the sense that it was the unique language of our analytic encounter (not even English, but our own invention, so to speak) and, thus, an embodiment of our "transferential chimera."[6] It was the membrane that resonated to the minute changes of our unconscious communications—sometimes becoming too rigid, sterile, and hardly understandable, sometimes dull and flat, sometimes used as a weapon in a combat of opposites, sometimes pulsing in an agony of lack of expression, sometimes running smoothly like a stream, and sometimes rotating in a search for eloquence, thus indicating that here is a deep whirlpool. So the foreign language played the role of a container that bridged us on a conscious level and marked a boundary at the unconscious level. Without it, I would have been seized by a vortex of a powerful incestuous drive which at that moment would have been too destructive for me. It is interesting to note here that in spite of the fact that my analyst was of a different gender, the transference was impregnated not only with the incestuous drive towards the "father" but also (or even mainly) with the incestuous drive of a matriarchal kind, one of the versions of the death impulse that are described in two articles by Martin-Vallas regarding "transferential chimera."[7]

But why can we not say the same about the mother-tongue? Is it not the integral part of the containing ability of an analyst, doesn't it belong to the analytic field? Isn't it also a sonorous representation of a *vas hermeticum* and subject to all possible individual differences? This is true, but somehow, in those circumstances when a person prefers (not necessarily fully consciously) a foreign language to the natal one, there is the meaningful ingredient of a negation of the mother, or maybe even a negation of a parental couple, caused by some deep disturbance or lack (absence) of some kind, most likely at an early stage of development.

Thus, my second hypothesis is that a foreign language plays the role of a symbolic father and is highly relevant in cases where there is a constellation of a negative mother complex, in situations where the two faces of the father—the imaginary or symbolic one and the oedipal one—are not well balanced in a personal history.[8]

The symbolic father as the other, the third, stretches out a hand, builds a bridge to cross troubled water, and becomes that Noah who happened to rescue living creatures by migrating to other lands. To continue the metaphor, we may suggest that a foreign language becomes a representation of the psychic forces that attempt to rescue an ego from annihilation, gives a chance of deliverance from an overwhelming psychic flood, and strives to transcend a deeply rooted trauma of some kind.

Hope of Deliverance from the Darkness That Surrounds Us

I want to illustrate what I am describing with a vignette from work with my patient whom I will call Mario. At his first session he surprised me by his business-like approach to the meeting—he took out a list with an agenda and began to read out the issues that upset him. Possibly, this foundation of a paper with text written on it gave him some confidence, something to lean on that reduced anxiety—thus, he would not forget to mention anything valuable.

His mother-tongue belongs to a Romance group but he speaks Swedish and English fluently and is learning Russian. We had analysis in English. His problems in personal relationships were the main issue he brought initially. He was not able to build a reliable relationship with one woman and he compensated this lack by maintaining several parallel relationships, not satisfying any. He felt that he was not understood and that something truly meaningful remained undiscovered. He increased the number of relationships but still some important part of his soul remained *incommunicado*.

A frequent image that appeared in my mind during my work with him was an image of a merry-go-round. The ponies were whirling around the centre, but no matter how fast they ran, there was no motion forward. It was a carousel which Mario could not get off; he chased after excitement, but any satisfaction escaped him. All his attempts to change and to obtain that elusive "something" led to nothing. The flashing pictures and the bright lights of this carousel ended in ennui and fear—what if it never ends?

Usually he talked quickly and a lot, trying to be gentle and nice. My attempts to go deeper seemingly hit a misunderstanding, some obstacle—as if unconsciously he was avoiding it, as if something was not letting us in, keeping us at a rational, intellectual level that leaned on (or hid behind) the language. Capacious, structured, laconic English became a reliable construction that supported his ego, the float that allowed him not to sink into the infinite and threatening chaos. On the one hand, this construction was a rescuing one (the scaffolding that supported the building of his identity), on the other, it was a psychic defense that did not allow him to get in touch with pain, panic, despair.

In his sessions I often started to confuse English grammar; I forgot the words and took a long time to reflect upon the question I wanted to ask—would it be sufficiently comprehensible? Or I tended to give too straightforward an interpretation. It seemed that my behavior reflected some part of his inner hopelessness and confusion and also his attempts to depend on rationality. From time to time he stopped talking and watched me directly and continuously as if he was clutching at a straw, as if he was afraid of falling down. And at these moments no language was needed—he just needed my eyes to hold him. Just as often, though, his mind flew very far away.

In childhood and adolescence he was very insecure. He had a hyper-anxious and over-protective mother and a depressive father. His anamnesis is full of traumas. His current life is socially quite safe but he suffers from a lack of meaning. Often he tells me that the meaning lies in a parallel world, and that we'll all arrive at the same end in this one—death. He shares with me his impressions of Russian culture that frequently surprise me a lot. I picture to myself quite vivid images of his everyday life. I imagine his girlfriends and his child, his parents, the *Mise-en-scènes* of his youth as if I am watching a movie. It is my way to finish building something that is lacking in the language, and my curiosity about the other culture. And also it is a countertransference sign—a vivid, warm, bright, and intriguing part of him that is split off and inaccessible. Most likely it refers to a pre-verbal stage, to a state of maternal reverie.

He learned English when he was already an adult—it was his compensation for what he lacked in adolescence, when, due to his parents' choice, he found himself in a college that was irrelevant to his interests. So the English language became a bridge that connected the hopeless world of his father's choice with the bright spectrum of possibilities that he was picturing to himself in his fantasies which the outer world offered him and he strove to find (and from which his mother protected him for so long, keeping him away from it, keeping him just for herself). At the same time the English language is the bridge over the abyss that separates him from his self, a bridge that we have to use and that keeps us on a level of "adult," rational communication.

Probably in his very early life he had experienced an overwhelming failure of nurture that forced his psyche to come back again and again and to start again and again. Probably, he was re-enacting the original despair of the impossibility of any rapport with his mother, when he was unable to express his feelings and had no support from his father. He has almost no memories of his childhood and very few dreams, so there is a lot to reconstruct.

Sometimes our sessions tend to look as if we are just chattering (thus compensating for the lack of space for friendly talks in his private life) in this lighthearted, nice, intimate space—that compassionate field between equals, where sharing emotions and opinions is possible, that space of trust that he values so much—where he would be welcome to voice his emotional experiences. Such friendly small talk and the moments when he shares his impressions of a movie or an art exposition soften his defenses, extenuate the mechanical contact, and deepen the process, leading us to little insights. At such moments I know that it makes no difference which language analysis is choosing; the most significant thing is the fact that it goes on.

Things Left Unsaid

The facets of a language's role in analysis are endless. And in cases where this language is a foreign one, its role is emphasized and acquires additional nuances and connotations. I have attempted to show some of these nuances—a foreign language as playing the role of a boundary keeper, a container, a pillar supporting the interior space of neutrality, a rescue vessel, and a womb for bearing the meaning not yet ready to emerge. But the most delightful thing is to observe the process of how the effort to be understood crosses the gap that separates two different consciousnesses and unites them.

The mental effort that we make in order to transmit meaning in a different language allows us (in spite of all confusions, obstacles, misunderstandings) to construct this bridge of understanding at each and every moment of communication, and more often than not we find the means to do it. Thus the transcendent function, so much talked about among Jungians, is being invited into action and helps trust and meaning to be born and gives the opportunity to connect us to each other and to our Selves, "without any words."

Personal Note

While I was writing this article I had a dream:

I saw my older foreign colleague. ... A man was presenting her with a ring. There was a figure of a young god on this ring; the figure looked to me like Hermes. Having put this ring on her finger my colleague laughed lightly and happily.

It was a happy dream, indeed—a dream in which I saw that my ability to act energetically (a man) was engaged to my intuitive wisdom (my colleague) in the name of and thanks to a communicative gift (Hermes). And this joyful union was a sign to me that the two halves of myself were being successfully united and that my long-term efforts to bridge these two inner realities were not in vain. I woke up in a peaceful and light mood. So writing had its healing effect.

Postscript: Silly Love Songs

The best way to learn a foreign language, especially for those who love music, is to listen to songs and sing them. The next step is to watch movies. And one just can't help loving the people whose language he is speaking and singing. For people of my generation, inhabitants of the "golden autumn" of the Soviet Union, those songs were the window to "freedom." Maybe that's why I have tried to use phrases from songs (and movies) as the sub-headings of my chapter:

"*Voyage voyage*," Desireless, 1987
"Bridge Over Troubled Water," Simon and Garfunkel, 1970
Lost in Translation, directed by Sophia Coppola, 2003
"Feelings," written by Louis Gasté, performed by Morris Albert, 1974
Analyze This and *Analyze That*, directed by Harold Ramis, 1999 and 2002
"People Are Strange," The Doors, *Strange Days*, 1967
"Hope of Deliverance," Paul McCartney, *Off the Ground*, 1993
"Things Left Unsaid," Pink Floyd, first composition, *The Endless River*, 2014
"Silly Love Songs," Paul McCartney, *Wings*, 1976

Notes

1. Elizabeth Gordon, "Three in the Room" (lecture, 2nd European Conference for Analytical Psychology, St. Petersburg, Aug.–Sept. 2012).

2. Nathan Schwartz-Salant, *Narcissism and Character Transformation*, trans. Valery Mershavka (Moscow: Class, 2007).

3. Kenneth Reich, "From Object Relations to the Theory of Relations: Hope in Couples' Therapy" (lecture, Russian-American Conference on Psychotherapy, Moscow, Oct. 2005).

4. Jacqueline Gerson, "Malinchismo: Betraying One's Own," in Thomas Singer and Samuel L. Kimbles, eds., *The Cultural Complex: Contemporary Jungian Perspectives on Psyche and Society* (New York: Brunner-Routledge, 2004).

5. Pierre Benghozi, "*Anamorphoses, mue des contenants et transformations psychiques familiales*" (lecture, European Federation for Psychoanalytic Psychotherapy [EFPP] Conference, Florence, 2010).

6. Francois Martin-Vallas, "The Transferential Chimera: A Clinical Approach," *Journal of Analytical Psychology* 51 (5, 2006): 627–41.

7. Francois Martin-Vallas, "The Transferential Chimera II: Some Theoretical Considerations," *Journal of Analytical Psychology* 53 (1, 2008): 37–59.

8. Julia Kristeva, "*Soleil noir: depression et melancolie*," trans. D. Kralechkin (Moscow: Cogito-Center, 2010).

Chapter 11

The Delivery of Training
Personal Experiences as a Trainer in Other Cultures

Angela M. Connolly

Introduction

Over the last few decades we have witnessed an increasingly profound crisis in psychoanalytical training in the West, a crisis that has manifested itself on the one hand by the fall in number and the advanced age of candidates, and on the other, by the increasing feeling among trainers that something seems to have gone wrong in the way in which we transfer our theories and our clinical skills to future analysts. The same is certainly not true, however, in countries and regions such as Russia and Asia where depth psychology was relatively unknown until recently and where the creation of Jungian training programs has usually been greeted with enormous enthusiasm and whole-hearted participation, at least in the initial stages. I have been fortunate enough to have had the privilege of working first as the personal analyst for candidates in Moscow where I lived for five years, and then as a trainer in four very different cultural and linguistic contexts: as an International Association for Analytical Psychology (IAAP) examiner in the St. Petersburg program and as a teacher, supervisor, and examiner in the People's Republic of China, in Taiwan and in Kazakhstan.

The task of any trainer is both to impart the theoretical and clinical knowledge required to develop the skills and competences necessary for analytical work and at the same time evaluate the capacity of the candidate to develop these skills and competences. If this is no easy task working in our own institutes, it becomes even more complicated in areas where the culture, language, and political structure are very different from that to which we are habituated. Kim, in a paper of 2000, defines culture as "an emergent property of individuals interacting with their natural and human environments," a definition that allows us to appreciate that, given the nature of emergence where small initial differences lead to very large differences in the final result, indigenous cultures will inevitably be very different from our own.[1] Adapting our models to different cultural contexts has therefore required considerable effort, much flexibility and patience, and no little capacity to become aware of our own preconceptions and to invent new strategies to resolve unexpected problems. Nevertheless, these experiences have been both enriching and thought-provoking, and, for me, they have been a fruitful source of stimuli for a potential revision of our own more traditional training models. I am grateful to the editors for providing me with this opportunity to look back and reflect on my experiences and their personal and collective significance.

The Aims of Training

As I noted in a previous paper,

> Analytical training has as its fundamental scope the transmission of skills: phenomenological skills in the sense of being able to observe as clearly and as accurately as possible what goes on in the analytical setting; hermeneutical skills in the sense of being able to make interpretations that are faithful to the clinical facts observed and are capable of creating new meanings; theoretical skills in the sense of being able to reflect on clinical facts and to measure them against pre-existing knowledge.[2]

As well as all these skills I listed then, there are, however, two other skills which are rather more difficult to teach and to develop but which are fundamental to the practice of analysis. One of those skills is, I believe, common to all analytical schools but the second is specifically linked to the Jungian method. From Bion onwards, analytical thinking has increasingly stressed that the ability to make *good* interpretations (and I am using this term in the widest sense possible) requires both the capacity to think metaphorically and the capacity to use reverie. As Warren Colman says,

> The more metaphors analysts have available and the more adept they are at using them, the more skilfully imaginative they are likely to be at formulating vivid interpretations that speak directly to the patient's material in metaphorical and/or symbolic terms.[3]

It is not sufficient, however, to be able to think metaphorically. A good interpretation requires metaphors that are both appropriate and capable of adding depth to the patient's material. When we think metaphorically we will for the most part be making use of the repertoire of conventional metaphors that belong to the particular culture in which we work, but an analyst must also be capable of inventing new and original metaphors.

New metaphors, as Lakoff and Turner point out do not depend so much on the capacity to invent something new *tout court* but rather on the ability to combine conventional metaphors in an original way.[4] Original metaphors work not only to help us understand our psychic processes and structures, they also possess the ability to add something new to our experience "through the power of their expressive capacity to give form to unprocessed experiences, their ability to link together different experiences and their capacity to hold the creative tension between thought and sensation, feeling and intuition."[5] This is a fundamental characteristic of interpretations based on reverie which spring from a spontaneous image that is created in the mind. As Antonino Ferro writes, reverie is more similar to "a fragment of a dream, a visual pictogram, fruit of that waking dream, thinking that is continually taking place."[6] All this is in line with the two basic tools of the Jungian method, amplification and active imagination, both of which are aimed at adding metaphoric and symbolic depth to the images of the patient's dreams and fantasies. Reverie is particularly important in trauma where the representational capacity has failed and where it is the analyst who needs to be able to dream the patient's dreams.

The second skill that a successful analytical training should be able to teach and develop is a particular kind of analytical attitude that is specific to the Jungian method. George Bright in an article of 1997 has suggested that

> Analytical attitude in Jungian analysis is thus based on the concept that there is an underlying, psychoid dimension to meaning. We acknowledge this by taking everything that happens within analysis seriously but not literally. We acknowledge it too by our attitude of restraint towards explicitly stating the meaning of any aspect of the patient's experience—or, indeed, of our own experiences in the countertransference—as if our grasp of the meaning were absolute. This attitude of restraint is based on the premise that while there is an underlying meaning and patterns in all human experience, that it is unconscious, and can therefore only be known indirectly and in a provisional way.[7]

A correct analytical attitude therefore requires the capacity to tolerate uncertainty and to resist reaching out for premature understanding. As Jung says, "The danger of wanting to understand the meaning is over-valuation of the content which is subjected to intellectual analysis and interpretation, so that the essentially symbolic character is lost."[8] If the current models in use in our institutes are usually fairly adequate in imparting phenomenological, hermeneutic, and theoretical skills, where our traditional models tend to fail, however, is in the teaching of the last two skills, and it is exactly here that I believe we have much to learn from the experience of training in different cultures. What I propose is first to take a brief look at our traditional training models and the problems that have led to the crisis in training, then to look at some characteristics of Asian culture that I have found important, and finally to look at what I have learned from this encounter with "otherness" and the way in which it has influenced both my theories and practice and the way I theorize and practice training.

Training Models in the West

To ensure the transmission of all these different skills, various models of training have been put in place. Most of the models in use are actually derived from the so-called Eitingon model with its tripartite structure of training analysis, supervision, and theoretical and clinical seminars, but two alternative models of training have also been instituted: the so-called French model and the model of Kachele and Thoma.[9] The French model, described by Kernberg aimed at separating the training analysis from the rest of the training in order to reduce dependency and limit the real influence of the training analyst on the life of the analysand.[10] Yet another model has been put forward by Kachele and Thoma in their 1998 memorandum on psychological education where they suggest that the Eitingon tripartite model is really a degeneration of the original model as conceived of by Freud and Eitingon when they set up in the Berlin Poliklinik as a research and teaching institute offering free psychotherapy for the masses.[11] Kachele and Thoma stress the need to re-introduce the classical triad of teaching, treatment, and research in order to combat excessive authoritarianism and to revivify psychoanalytical training and knowledge. They stress the need to preserve the autonomy of the personal analysis and to limit the power of the institute over the length of the training analysis. As Thoma says,

> There is every indication that the present day crisis of psychoanalysis is an indirect consequence of a training system, which over the past 40 years or so, has ever more extended the length of training analysis and given it a central position in the training.[12]

Although the establishment of training institutes was long resisted by Jung because of his realization of the opposition between the institutional demands of training and the analytical aim of individuation, nevertheless, the increasing pressure to provide a professional and standardized training, necessary if analytical psychology was to be able to survive and compete with other schools of depth psychology, led him to acquiesce in the setting up of the first training institutes. In both the Society of Analytical Psychology founded in London in 1946 and the C. G. Jung Institute of Zürich founded in 1947, the basic model of training was once again that of Eitingon, "personal analysis, study, practical casework"[13] and the same problems of authoritarianism, theoretical dogmatism, and a power differential between training analysts and candidates gradually made themselves felt. Today, many Jungian institutes have adopted some of the modifications described above in the attempt to improve training, and not a few have abolished the category of training analyst altogether. Nevertheless, despite the undoubtedly positive effects of more democratic processes and greater transparency in our organizational structures, we still have to recognize that there remains something in the way our institutes work that is in need of critical revision.

THE CRISIS IN TRAINING MODELS

Kernberg, in three provocative papers, attributes the crisis in training to the pernicious effects of the training offered by institutes whose organizational structures seem "to correspond best to the combination of a technical school and a theological seminary"[14] in which, all too often, the main objective seems to be that of acquiring "well-proven knowledge regarding psychoanalysis to avoid its dilution, distortion, deterioration, rather than helping students to acquire what is known in order to develop new knowledge."[15] Kernberg feels that this is linked in part to the role of training analysts, an institution that all too often degenerates into a kind of administrative oligarchy that exerts a rigid control over psychoanalytical institutes. Casement too links the problems in training to the authoritarian atmosphere of training institutes which makes many candidates feel infantilised during training.[16] Feedback from candidates about their training experience strengthens these hypotheses. Bruzzone et al. suggest that while there is an inevitable conflict in any training between the level of the personal analysis with its regression and its transference dynamics and the level of theoretical and clinical training which require a certain maturity, this conflict can be accentuated if there is an excessive attitude of protection and caution towards the candidate on the part of teachers and supervisors and a failure to accept full pedagogic responsibility.[17] All these authors link the difficulties experienced in training to the models of training utilized by the institutions. As LaCapra says,

> Institutions (in the largest sense) are normative modes of 'binding' with variable relationships to more ecstatic, sublime, or uncanny modes of 'unbinding.' They require repetitive performances which may become compulsive but may also facilitate exchanges.[18]

More and more trainers are being forced to recognize that it has now become necessary to make important changes in our training models, and it is exactly here that experiences in training in regions and countries where there are no institutes and where the cultural context is radically different can provide us with important insights.

Studying Jung in Asia

In this chapter I will be reflecting on some of the cultural differences and the problems which visiting teachers and supervisors find themselves facing when we have to adapt our Western theories and models of training to Asian cultures such as China or Taiwan. The re-introduction of psychology and analytical psychology into China and Taiwan is relatively recent. In China, during the Great Cultural Revolution which lasted from 1966–1968 (although its effects were still felt until 1971), all university psychology departments were shut down and were only reopened in 1976. Interest in all analytical schools of thought could then begin to be openly expressed and in the nineties, Heyong Shen, professor of Personality Psychology at the South China Normal University in Guangzhou, began to teach courses on Jung. The first conference on Analytical Psychology and Chinese Culture was held in 1998 and since then conferences have continued to be held both in China and Taiwan. Professor Shen became an individual member of the IAAP in 2004 and there are now four qualified analysts and Developing Groups have been recognized in Guangzhou, Hong Kong, Shanghai, Macau, and Beijing. In Taiwan in 2001 The Taiwan Institute of Psychotherapy was set up under the chair of Jenny Chang and the directorship of Dr. Hao-Wei Wang, and a Developing Group was recognized in 2010.

Inevitably, at the initial stages, the organizational structures and the processes put in place had a somewhat improvised ad-hoc quality with many inconsistencies and incoherencies. All this and the explosion of interest in Jungian analytical training all over the world led to a radical reorganization of the IAAP program under a new Education Committee with the aim of ensuring a coherent and containing organizational environment and adequate evaluation procedures. Almost twenty years have passed since the first contacts with China and the time is now ripe to begin to reflect not just on the assessment and evaluation of our training programs. It is also fundamental, however, that we begin to make use of our experiences teaching and supervising in different cultures as this can offer important insights and new directions for the future development of Jungian theory, clinical practice, and training. Working in other cultures forces us to become aware that we do not all think alike and this has important implications not only for training but also for the relationship between analytical psychology and other psychoanalytical schools.

Christopher Bollas in *China on the Mind* presents some thought-provoking reflections on his experiences in his contacts with Japan and Korea and on the way this has modified his thinking on psychoanalysis. Bollas distinguishes between what he terms the *paternal order* and the *maternal order*. The paternal order is dependent on language and the verbal. It stresses rules and clear lines of demarcation between self and other and favours self-achievements, conflict, and the development of the individual self. The maternal order is based on non-verbal forms of being, thinking, and relating; it stresses images rather that words and favours harmony and the development of the collective. The paternal order is representational in the sense that it stresses the content of communications and the narrative. The maternal order is presentational, stressing the form rather than the content and is essentially poetic. According to Bollas,

> The maternal order that is foundational to psychoanalysis has been subjected to an ongoing repression within the psychoanalytical movement but since this presents an Eastern way of being and relating, is it possible that growing commerce between West and East will de-repress the maternal order and challenge the hermeneutically-bound causal-inclined paternal focus that has dominated psychoanalytical discourse?[19]

In a certain sense the opposite is true of Jung's approach to analytical theory and practice which undoubtedly favoured the maternal order. Elizabeth Lloyd Mayer, in a 2002 paper, has suggested that while the Freudian model of mind emphasizes "its separateness and its unequivocally boundaried character," the Jungian model stresses a radically connected and unboundaried mind.[20] In the same way, while the psychoanalytical method is based on words, the Jungian method places emphasis on the image. As George Hogensen says, "For Jung the visual image is as important as language is to Freud."[21]

Following the work of Fordham and the developmental school, however, many Jungians now have integrated models of practice and training that place a much greater emphasis on boundaries, frames, and on working in and with the transference, just as the emphasis on object relations and on a relational model of clinical practice has further modified our theories and practice.[22] All this has meant that we have sometimes tended to focus less on some of the essentials of the Jungian model of practice and theory. Coming into contact with the Eastern mentality, which has so many features in common with Jung's theoretical and clinical model, has helped me reconsider the importance of certain key concepts of Jung and has allowed me to begin a re-evaluation of how we train in the West.

Training in Different Cultural Contexts

If the task of any trainer working in our own Western institutes is not an easy one, then in the encounter with very different cultures the trainer is also faced with the problem that this encounter, if accepted, inevitably changes one's way of thinking and perceiving. The risk in this rather unsettling and destructuring experience is that the trainer will cling obsessively to her own cultural identity, her own models of training, and her own theories and methods and refuse to allow the culture to get under her skin, to permeate her, to transform her, her theories, and her method. If, as Jung put it, the analytical situation implies that the analyst be infected with the patient's illness (and this is also true of supervision, although to a lesser degree perhaps) how much more dangerous is it when we risk becoming infected with the virus of "otherness?"[23] Nevertheless, such experiences can lead to transformation and I believe a greater analytical flexibility and an increased capacity to listen to and work with the unconscious. As Julia Kristeva writes,

> Those who have never lost the slightest root seem to you unable to understand any word liable to temper their point of view ... the ear is receptive to conflict only if the body looses its footing. A certain imbalance is necessary, a swaying over some abyss, for a conflict to be heard.[24]

From my own personal experiences living and working in Russia, I am convinced that it is exactly this experience of *loosing* my roots and having to come into contact with the lack of boundaries and the collective mentality of post-Soviet Russia that has facilitated my coming-to-terms-with the extreme "otherness" of cultures such as those of Taiwan and China. Working in Asian cultures has inevitably changed the way in which I think and work as an analyst, emphasizing the value of certain of our theoretical concepts and constructions, our clinical method, and our downplaying or modifying others. What I propose now is to look at two essential concepts which are foundational in Asian culture in order to trace out the way in which they both affirm and modify certain key concepts in Jungian theory, practice, and training: *mianzi*, or *face*, and its relevance to the role of the personal equation in analysis and *yuan*, or the idea of the importance of collective harmony, and its influence on the construction of the Self and individuation. I will then look at the relationship between Eastern states of mind and Jung's approach to theory, and finally I will look at the different ways of approaching the unconscious and the importance of the image in Asian culture and how they confirm the importance of the Jungian clinical approach with its attraction to image.

Mianzi, the Personal Equation and Jungian Practice

Ruth Benedict in her ground-breaking 1946 study of Japanese culture, *The Chrysanthemum and the Sword*, put forward the hypothesis that while Western societies are fundamentally guilt cultures, Japan is a shame culture.[25] This is equally true of other Asian cultures all of which stress the importance of *mianzi*, (face, giving face, showing respect). This concept regulates the interactions with superiors and authority figures but even more importantly it sets forward the acceptable ways of self-presentation which are very different from the Western style of self-presentation through words and narrative. Megumi Yama, in a 2013 article, conveys the difference between these two different approaches in the description she gives of her difficulty in readapting to Eastern styles of communication after spending her adolescence in America. "I was careful not to express myself clearly and directly as this would run counter to the Japanese style of communication where we are expected to express ourselves in an implicit way and guess the intentions of the other from a given context."[26]

The concept of *mianzi* also has important implications for how we train and supervise in Asia. In Eastern cultures what counts is not so much what one says but the way in which one says it and the whole style of self-presentation. In the words of Lao Tzu, "Wisdom resides not in what one says, not in what one accumulates but in the way one lives."[27] This approach, with its contrast between explicit self-representation through words and implicit self-presentation is very similar to Jung's vision of therapy. For Jung, as he himself states in the clearest possible way, psychotherapy is "a dialectical procedure" in which "the therapist is no longer the agent of treatment but a fellow participant in a process of individual development."[28] Hence, in Jungian analysis, it is the personality of the doctor which is the fundamental factor in analysis, for, as Jung writes, "we have learned to place in the foreground the personality of the doctor himself as a curative or harmful factor."[29] Again, "the personalities of doctor and patient are often infinitely more important for the outcome of the patient than what the doctor says and thinks (although what he says and thinks may be a disturbing or healing factor not to be underestimated)."[30] If these injunctions of Jung are a fundamental part of any Jungian therapy, then the place where I think we tend to forget the importance of self-presentation and the personality of the analyst is in training. All too often, when we train, the emphasis is on what the trainer says rather than on who the trainer is, on his or her total personality and personal equation. If a fundamental part of any Jungian analysis is the emphasis on individuation and on the need to respect and develop the personal equation of the patient, surely this is of equal, if not paramount, importance in imparting our theories and clinical methods.

YUAN AND INDIVIDUATION

As Toshio Kawai notes, "seen culturally and historically, modern consciousness is an achievement in Western history," an achievement that can be linked on the one hand to the establishment of the observer in the use of perspective and, on the other, to the establishment of private space.[31] This development of modern consciousness was linked to a drastic change in the way in which subjectivity, the sense of the *I*, was located and constructed. As Kawai notes, "the subjectivity that was formerly located in nature, community and other phenomena, came over time to be seen to be located in each individual human being."[32] The formation of the sense of a self located within the confines of the individual body and mind thus became based on processes of separation and differentiation, first from the mother, then from the family, and finally from the community—a process that was fundamentally linked to the struggle between the conflicting and opposing desires for separation and for union. The maintenance of the continuity of subjectivity that formerly depended on the community and on social ritual, now came to depend on the capacity, through language, to construct narratives capable of representing the self. Psychoanalysis itself, the talking cure, was born out of the need of individuals who were unable, because of their inability to resolve conflicts, to maintain this stable sense of personal identity.

In Eastern culture, on the other hand, there are fundamental differences in the way in which the sense of self and the relationships with the collective and with other individuals are constructed. Markus and Kitayama stress the opposition between the independent self and the interdependent self, between individualism and collectivism, a structure that is common to all Asian societies which are based on Confucian ethics.[33] Confucian ethics delineate eight principles of moral behaviour: loyalty, respect, kindness, love, trust, justice, harmony, and peace. These principles are the foundation of Chinese human relationships and networks and lie at the basis of the Chinese concept of *yuan* which Yang and Ho describe as, "a concept of interpersonal relationships that helps to maintain interpersonal harmony and group solidarity."[34] As Kakar notes, the Confucian tradition stresses

> an idea of mental health and psychological maturity where the self is appropriately responsive and in tune with the situations and persons of daily life: family, friends, colleagues at work, without an artificial boundary that shrinks the existence of the self to an individual unit.[35]

Jung's definition of individuation, at first glance, appears to favour the individual and his/her self-realization over the collective well-being. As he writes, individuation

> means becoming an 'in-dividual,' and so far as 'individuality' embraces our innermost, last and incomparable uniqueness, it also implies becoming one's own self. We could therefore translate individuation as 'coming to selfhood' or 'self-realization.'[36]

Jung is rightly suspicious of a collective harmony which is preserved at the cost of the development of the individual, and indeed the dangers of such an attitude are all too present in societies such as the USSR and in part also in China. In his later alchemical writings, however, and especially in *Mysterium Coniunctionis*, he begins to open up to the idea that the final stages of the alchemical process and of individuation imply a union with the world, an entering-into-contact with the *unus mundus*. The world as *unus mundus* is not, however, as he says, "the world of multiplicity as we see it but with a potential world, the eternal Ground of all empirical being, just as the self is the ground and origin of the individual personality, past, present and future."[37] This is very close to the idea of Dao, which in Chinese culture has always provided a counterbalance to Confucianism in order to produce equilibrium between the needs of the individual and those of the social group. If we think of the conflicts and splits that have bedevilled our training institutes in the West, perhaps we have something to learn from the Confucian tradition.

Eastern States of Mind and Jungian Theory

One of the most unsettling insights that derive from any encounter with Asian culture is the realization that their way of thinking is actually different from our own. Bollas, in his analysis of the different ways of thinking that are characteristic of East and West, suggests that, historically, Eastern thinking has tended towards forms of thought that are based on the maternal order, while Western thinking reflects forms of thought derived from the paternal order. For Bollas, "The Eastern mind uses language to create possible interpretations of meaning and is implicit rather than explicit. The Western mind seeks lucid definitions that are explicit and not meant to be open to the other's readings."[38]

If Freud's theory is essentially conceptual and his method language-based, this is certainly not true of Jung. For Jung, "One could as little catch the psyche in a theory as one could catch the world. Theories are not articles of faith, they are either instruments of knowledge and of therapy, or they are no good at all."[39] This relativization of conceptual thinking and theories is clearly manifested in Jung's writings. As Susan Rowland puts it, "Jung's writing on concepts is dialogical. He weaves a text-ure of meaning-making relationship between the intrinsic comprehensibility of logos concepts and the *vitality* of living mystery through stories that touch the unknowable creative psyche."[40] Teaching theory requires that we respect, and that we teach our students to respect, the polysemantic richness and ambiguity of his texts and the fact that so many of his ideas are expressed not only (or not at all) in terms of *either/or*, or of *true/false*, but rather in terms of *both/and*. Analytical theories are not conceptual nor do they represent absolute truths. As Warren Colman writes, "They are like metaphorical maps to the ever-shifting territory of the psyche, subject to the idiosyncratic descriptions of the mapmakers and only roughly applicable to the particular psychic territory which the analyst is likely to meet."[41]

Eastern Approaches to the Unconscious, the Image and the Psychoid

Hayao Kawai, in a 1976 work, suggested that in Japan the boundary between consciousness and unconsciousness is much vaguer than that which we find in Western individuals and that while the ego complex is the centre of consciousness in the West, in the East it is the Self which is the centre of consciousness.[42] In the West, the development of consciousness leads, as Toshio Kawai notes, to a substantiation of ego and unconscious and thus to a radical separation between ego and Self; in the East, on the other hand, the "distinction between ego and self is not important and is more fluid."[43] This means that the approach to the unconscious requires not so much a direct approach through words and linear thinking but rather the establishment of connections with the self through images which are arrived at through more circular routes such as dreams, painting, or sandplay rather than, say, interpretations of transference and countertransference. In the Eastern approach to the unconscious, it is fundamental to allow the images to unfold spontaneously, avoiding any anxious reaching-out for understanding and premature interpretations which all too often spring from the analytical superego rather than from the needs of the analysand. What training needs to insist on is how to help the candidate achieve the capacity "to allow the images to begin to emerge, express, and unfold themselves and to refrain from pushing for verbal articulation before what has been invisible is ready to be made manifest."[44] This is in line with the later developments in Jung's thoughts, his ideas on the psychoid unconscious and his insistence on the dangers of understanding. As George Bright states so forcefully, "Jung's formulation of the psychoid unconscious effectively works against the destructive effects of premature or over-definite interpretive understanding."[45]

Conclusion, or "Bringing It All Back Home"

In this present work I have tried to briefly describe some of the ways in which the encounter with the "otherness" of Asian culture has modified how I imagine and practice training through re-evaluation of some of the fundamental characteristics of Jungian theory and method. It might seem that I have taken an excessively rosy view, and I certainly do not wish to ignore the enormous level of collective trauma which I have already discussed in previous papers just as I am all too aware of how an excessive insistence on Confucian tradition has hampered the individual development of women in Asia.[46] Nevertheless, what I wish to underline here, are the benefits of the opening up to "otherness." As the anthropologist Clifford Geertz says,

> It is the asymmetries ... between what we believe or feel and what others do, that makes it possible to locate where we now are in the world, how it feels to be there, and where we might or might not want to go. To obscure those gaps and those asymmetries by relegating them to a realm of repressible or ignorable difference, mere unlikeness ... is to cut us off from such knowledge and such possibility: the possibility of quite literally, and quite thoroughly, changing our minds.[47]

If there is a profound disenchantment in the West about our models of training, the encounter with the East offers us the possibility of revitalizing our approach to training. I believe that when we train we need to be able to impart theoretical knowledge that stresses the non-conceptual, indefinite, and metaphorical nature of analytical theory just as we need to be able to help our candidates understand that good analytical work requires the capacity to tolerate waiting and not-knowing in order to be able to offer images and interpretations that are unsaturated, metaphorical, and original, capable of opening up to the inexhaustible symbolical richness of the psyche.

I have learned many things from the East, but perhaps the most fundamental lesson is the importance of the image and of the aesthetic in theory and practice, and I wish to conclude with the words of two Italian psychoanalysts, Domenico Chianese and Andreina Fontana in their 2010 book *Immaginando*. They advocate the need in psychoanalysis for an "iconic turn," capable of bringing into play different levels and meanings:

> the meaning we attribute to the dream (which remains a 'visual experience,' an 'aesthetic experience,' which can only partially be reduced to discourse), the significance we attribute to the unconscious (which is reductive to think of in terms of a language in as much as it has an essential and foundational relationship with the visual), [and] the significance we give to the field in therapy which we need to think of not only, and not so much, as a field of words but rather as the field of the image.[48]

Finally, I wish to take the opportunity of thanking my colleagues from China, Taiwan, and Japan for their unfailing hospitality and patience and for the way in which they have helped me to open up to new ways of imagining and practicing analysis and training.

NOTES

1. Uichol Kim, "Indigenous, Cultural, and Cross-Cultural Psychology: Theoretical, Philosophical, and Epistemological Analysis," *Asian Journal of Social Psychology* 3 (3, 2000): 265–87.

2. Angela Mary Connolly, "Some Brief Considerations on the Relationship between Theory and Practice," *Journal of Analytical Psychology* 53 (4, 2008): 481–501.

3. Warren Colman, "Theory as Metaphor: Clinical Knowledge and Its Communication," *Journal of Analytical Psychology* 54 (2, 2009): 199–217.

4. George Lakoff and Mark Turner, *More Than Cool Reason: A Field Guide to Poetic Metaphor* (Chicago, IL: Chicago University Press, 1989).

5. Angela Mary Connolly, "The Cognitive Aesthetics of Dream Metaphors: Some Reflections on Dream Work and Non-Dreaming in Psychosis and Perversion," in ed. Angiola Iapoce, *Understanding Jung* (Roma: Fattore Umano Edizioni, 2013).

6. Antonino Ferro, *Le viscere della mente: Sillabario emotivo e narrazioni* (Milano: Raffaello Cortina Editore, 2014), p. 43.

7. George Bright, "Synchronicity as a Basis of Analytical Attitude," *Journal of Analytical Psychology* 42 (4, 1997): 613–35.

8. C. G. Jung, "The Transcendent Function" (1916), in *Structure and Dynamics of the Psyche*, vol. 8, *The Collected Works of C. G. Jung*, ed. and trans. Gerhard Adler and R. F. C. Hull (London: Routledge & Kegan Paul, 1960), § 176.

9. Max Eitingon, "Report of the Berlin Psycho-Analytical Policlinic," *International Journal of Psychoanalysis* 4 (1923): 254–69.

10. Otto Kernberg, "A Concerned Critique of Psychoanalytical Education," *International Journal of Psychoanalysis* 81 (2000): 97–120.

11. Horst Kachele and Helmut Thoma, *Memorandum about a Reform of the Psychoanalytic Education* (unpublished manuscript, 1998).

12. Helmut Thoma, "Training Analysis and Psychoanalytic Education," *Annual of Psychoanalysis* 21 (1993): 3–75.

13. James Hillman, "Symposium on Training II: Training in the C. G. Jung Institute," *Journal of Analytical Psychology* 7 (1, 1962): 3–19.

14. Otto Kernberg, "Changes in the Nature of Psychoanalytic Training," in ed. Robert S. Wallerstein, *Changes in Analysts and in Their Training* (London and New York: International Psycho-Analytic Association, Monograph 4, 1984), pp. 56–61.

15. Otto Kernberg, "Thirty Methods to Destroy the Creativity of Candidates," *International Journal of Psychoanalysis* 77 (8, 1976): 1031–40.

16. Patrick Casement, "The Emperor's New Clothes: Some Serious Problems in Psychoanalytic Training," *International Journal of Psychoanalysis* 86 (4, 2005): 1143–60.

17. M. Bruzzone, E. Casuala, J. P. Jimenez, and J. F. Jordan, "Regression and Persecution in Analytic Training: Reflections on Experience," *International Review of Psychoanalysis* 12 (1985): 411–15.

18. Dominick LaCapra, "History and Psychoanalysis," in ed. Françoise Meltzer, *The Trial(s) of Psychoanalysis* (Chicago & London: University of Chicago Press, 1987), p. 31.

19. Christopher Bollas, *La Mente Orientale: Psicoanalisi e Cina* (Milano: Raffaello Cortina Editore, 2013), p. 29.

20. Elizabeth Lloyd Mayer, "Freud and Jung: The Boundaried Mind and the Radically Connected Mind," *Journal of Analytical Psychology* 47 (1, 2002): 91–99.

21. George Hogenson, "Archetypes as Action Patterns," *Journal of Analytical Psychology* 54 (3, 2009): 325–39.

22. Michael Fordham, *Technique in Jungian Analysis* (London: Heinemann, 1974).

23. Jung, "Principles of Practical Psychotherapy" (1935), in *The Practice of Psychotherapy*, CW 16, § 23.

24. Julia Kristeva, *Strangers to Ourselves*, trans. Leon Roudiez (New York, London, Toronto, Sydney, Tokyo, Singapore: Harvester Wheatsheaf, 1991), p. 17.

25. Ruth Benedict, *The Chrysanthemum and the Sword: Patterns of Japanese Culture* (Cleveland and New York: Meridian Books, 1946/1967).

26. Megumi Yama, "Ego Consciousness in the Japanese Psyche: Culture, Myth, and Disaster," *Journal of Analytical Psychology* 58 (1, 2013): 52–73.

27. Lao Tzu, *Dao De Jing: "Making This Life Significant," A Philosophical Translation*, trans. R. T. Ames and D. I. Hall (New York: Ballantine, 2003), chap. 81.

28. Jung, CW 16, § 7.

29. Jung, CW 16, § 172.

30. Jung, CW 16, § 163.

31. Toshio Kawai, "Postmodern Consciousness in Psychotherapy," *Journal of Analytical Psychology* 51 (3, 2006): 437–51.

32. *Ibid.*

33. Hazel Rose Markus and Shinobu Kitayama, "Culture and the Self: Implications for Cognition, Emotion, and Motivation," *Psychological Review* 98 (2, 1991): 224–53.

34. Kuo-Shu Yang and David Ho, "The Role of Yuan in Chinese Social Life: A Conceptual and Empirical Analysis," in eds. A. C. Paranjpe, D. Y. F. Ho, and R. W. Rieber, *Asian Contributions to Psychology* (New York: Praeger, 1988), pp. 263–81.

35. Sudhir Kakar, "Psychoanalysis and Eastern Spiritual Healing Traditions," *Journal of Analytical Psychology* 48 (5, 2003): 659–679.

36. Jung, *The Relations between the Ego and the Unconscious*, CW 7, § 266.

37. Jung, *Mysterium Coniunctionis*, CW 14, § 760.

38. Bollas, *La Mente Orientale*, p. 14.

39. Jung, *Medicine and Psychotherapy*, CW 16, § 198.

40. Susan Rowland, "Ghost and Self: Jung's Paradigm Shift and a Response to Zinkin," *Journal of Analytical Psychology* 54 (5, 2009): 697–717.

41. Warren Colman, "Theory as Metaphor: Clinical Knowledge and Its Communication," *Journal of Analytical Psychology* 54 (2, 2009): 199–217.

42. Hayao Kawai, *Pathology of Japan as a Society of Maternity*, trans. Megumi Yama (Tokyo: Chuokoron-sha, 1976).

43. Toshio Kawai, "Union and Separation in the Therapy of Pervasive Developmental Disorders and ADHD," *Journal of Analytical Psychology* 54 (5, 2009): 659–77.

44. Yama, "Ego Consciousness in the Japanese Psyche," pp. 52–73.

45. George Bright, "Jung's Concept of Psychoid Unconscious: A Clinician's View," in eds. Alessandra Cavalli, Lucinda Hawkins, and Martha Stevens, *Transformation: Jung's Legacy and Clinical Work Today* (London: Karnac Books Ltd., 2014), p. 104.

46. Angela Mary Connolly, "Through the Iron Curtain: Analytical Space in Post-Soviet Russia," *Journal of Analytical Psychology* 51 (2, 2006): 173–91.

47. Clifford Geertz, "The Uses of Diversity," *Michigan Quarterly Review* 25 (1, 1986): 114.

48. Domenico Chianese and Andreina Fontana, *Immaginando* (Milano: FrancoAngeli, 2010), p. 147.

CHAPTER 12

SHUTTLE ANALYSIS ACROSS CULTURES

PERSONAL ANALYSIS BY SHUTTLE: CAN IT WORK?
CHRISTOPHER PERRY

THE "FRONTIER-LINE IN HUMAN CLOSENESS"
ANN FODEN

PERSONAL ANALYSIS BY SHUTTLE: CAN IT WORK?
BY CHRISTOPHER PERRY

INTRODUCTION

This chapter is an attempt to evaluate the therapeutic usefulness of shuttle analysis in Russia with a relatively small number of analysands, some of whom had had previous experiences of psychotherapy/analysis.

Between 2002 and 2010, I flew four times per annum from London to St. Petersburg, Russia, as a member of a team of Jungian analysts, supervisors, and teachers who formed the Russian Revival Project, an International Association for Analytical Psychology (IAAP) enterprise founded by London analysts to fill the serious lacuna of training in analytical psychology that had been created by the rise of Communism.

Shuttle analysis consisted of meeting with each analysand for five ninety-minute sessions over a five-day period. This (potentially) intense engagement was then followed by a three-month fallow period, during which there was no contact. After my first four visits, I ceased working in an institutional setting because of my boundaries being disturbed by people and noise intruding into the analytic space, and I moved to a flat.

Context

The IAAP's aim was underpinned by a collective preconscious, if not unconscious, longing for analytic understanding in order to heal the personal and collective trauma inflicted by revolution, famine, war, the Gulag, and mass disappearances amongst which dwelt, like a ghost, the possibility of asking questions, which are the children of freedom. The totalitarian regime, recalled with nostalgia by some of my Russian friends, had done everything it could to impose a pall of amnesiac fog over the twentieth century so that there was little evident memorial to the millions who had died. Personal dissociation within individuals, many of whom had suffered appallingly, was underpinned by a collective dissociation enhanced by the excitement and manic defence of perestroika and later fuelled by a new leader who wanted to rehabilitate Stalin.

In psychoanalytic literature, totalitarianism has become enshrined in the concept of the "totalitarian object"—"a primitive pre-genital object … whose total power is intrusive and presupposes a total compliance and identification. As an introject … it is infectious and can survive in subsequent generations."[1] Beyond that, I think its infectious potential can extend to shuttle analysts, whereby the perceived "authoritarianism" of shuttle analysis by the analysands undermines their own and their analysts' authority.

To illustrate the enormous impact on me of this challenging context, let me share a dream I had during my first night in St. Petersburg. It was November, and the city was under a blanket of snow. As I went to sleep, I heard gunfire.

> *A starving, bedraggled urchin, about eight years old, led me by the hand below the River Neva into a vast dimly lit, cavernous space. We walked slowly along an enormously wide corridor,*

on either side of which were compartmentalised halls housing the dispossessed, the insane, the criminal; hordes of street children; spectres of thousands in exile; people transported to the Gulag; and the victims—alive and dead—of the siege of Leningrad {St. Petersburg}. I was overwhelmed by distress, and simultaneously in awe of my young guide, who seemed not only familiar with but also completely dissociated from this corridor of human tragedy and horror, along which we were slowly making our way. Once out of the tunnel, we had to climb a steep hill, on top of which we found ourselves in an old Russian wooden house. An ancient, timeless couple were sitting at the table, eating soup in candlelight and clearly awaiting our arrival. We were invited to join them in preparation for our continuing journey.

This was and is my dream. And yet, I still feel sure that it presented itself at the interface between my personal and Russia's cultural complexes. Jung wrote, "Almost every great country has its collective attitude. Sometimes you can catch it in a formula, sometimes it is more elusive, yet nonetheless it is indescribably present as a sort of atmosphere that pervades everything."[2] In addition to seeking to create connections with my analysands through their personal narratives, present pre-occupations and dreams of a future, we had the daunting task of finding moments of meeting between and across the Russian and English temperaments.

Working with an Interpreter

This "interactive field of strangeness" was both enhanced and relieved by working with and through my interpreter, who, very fortunately for me, was steeped in Russian culture, had a deep knowledge of all things Russian, and could explain to me areas of Russian life that differed significantly from our Western ways.[3]

My interpreter and I created together a container which included total confidentiality around each analysand as well as the time and space to debrief with one another about disturbing, shared countertransference experiences, translation misunderstandings, and cultural differences.

In addition to me having to adapt to a new way of working—having a third real person in the room—I needed to understand the analysands' projections onto my interpreter, which were initially quite negative and very uncomfortable for her; for example, she was seen as a *voyeuse*, an interfering mother-in-law, a servant, and a *whisperer*.

"The Russian language has two words for a 'whisperer'—one for somebody who whispers out of fear of being overheard, another for the person who informs or whispers behind people's backs to the authorities."[4] As trust developed, she faded into the background of the psychic space, where she was experienced as a benign and helpful presence.

Arranging the Analyses

Having negotiated entry with the female guardians of the East European Institute of Psychoanalysis—formidable, unsmiling ladies (my projection?)—I met my interpreter, who took me to a lecture theatre where I was to meet my first four analysands. Arranged triangularly on a rather small podium under the blackboard were three chairs. The podium was so small that I had an immediate anxiety in my stomach that any of us, through a slight shift backwards, could tilt onto the floor. This was an introduction to one of the myriad somatic countertransference sensations I experienced and had to befriend and understand as a trusty psychopomp into the relationship between analysand, interpreter, and myself.

As each of the four meetings drew to an end, a bell rang in the Institute signalling the end of lectures and the chance for hordes of students to "gate-crash" the lecture hall to collect their belongings. My interpreter was mystified by my consternation, even possibly a bit hurt. I had noticed that I seemed to be the only person disturbed by these invasions. She carefully explained to me that all of my analysands and, indeed, she had had to succumb to years of communal living, with very little lived experience of privacy. It became more and more apparent that my English understanding of boundaries would become conflicted with that of my analysands, some of whom lightly suggested that members of the Russian Revival Project were colonising missionaries!

Transference as Focus or Distraction

That sort of psychic banter scurried off into the ether as I became aware, with some alarm, of what I rather narcissistically misunderstood as resistance, but which I later understood as my imposition of a way of thinking rather than seeking authentic connection with each analysand. I am referring to my initial tilting towards transference interventions whereas what analysands primarily sought from me was an understanding of and relationship with their states of mind.

Here is a story that one told me: A man was sent into exile during the Stalinist purges. Three years later, he received a parcel from his Mama. It had been a tin box, but time and rust had reduced it to something resembling a sieve. Years later, when he was reunited with his Mother, she told him that it had contained some biscuits, tobacco, and a pair of warm woolly socks she had knitted for him. But—when he opened it, he was faced with a squidgy, useless conglomerate (something like the *massa confusa* of the Alchemists, I thought!). Nonetheless, he realised that he had not been forgotten—had not dropped out of the family's hearts.

This story was recounted during a January visit in my second year of work. I thought and verbalised (possibly out of guilt over the long break since October) that I was being told that: I was too late; what I had to offer was useless; that there was anxiety about being dropped out of my awareness; and that this was being defended against by an acknowledgement that I had not forgotten—because I had returned. Whilst some of this was reluctantly accepted and contributed to a slow-drying glue, bonding together the working alliance between the three parties in the room, the essential and eventually owned anxiety in the story (the analysand's fears about an inner rusting of psychic connections) was entirely missed by me.

With the help of my peer supervision group in London, I came to understand that this sort of archetypal amplification could act as my guide into the analysand's state of mind rather than towards attempts to work in the transference which, whilst important for me to acknowledge internally, was almost irrelevant to the analysand. Whilst the two are not mutually exclusive, experience increasingly supported a focussed, empathic exploration of the vicissitudes of various distressing and disturbing states of mind.

THE USE OF COUNTERTRANSFERENCE

Another guide was my musical countertransference, which informs much of my work, quite reliably. Deprived of natural linguistic understanding—like someone deprived of any of the five senses—and occasionally discombobulated by mistranslation (e.g., *authoritative* translated as *authoritarian*), I was helped by my musical countertransference in times of bewilderment, which surpassed the more familiar sense of "not knowing."

B had been "abandoned" by her previous shuttle analyst from London, who had to withdraw. When we first met, her dejection was palpable; it took us about a year of twenty double-sessions (ninety minutes) over four shuttle visits to develop some sense of trust and closeness. She had managed to evolve from relative self-isolating muteness to shared exploration of her psycho-somatic distress, which was the primary object of her curiosity.

She arrived for the first session of the fifth shuttle clearly not wanting to shake hands, averse to all contact, and reverted to muteness. As always, our interpreter sat motionless and unobtrusively, her hands folded on her lap, her gaze occasionally lifting from its resting place on the floor to glance at B, but without expectation.

As the silence settled into increasing unease, I started hearing a piece by Webern—a piece typical of his sparsity of notation and one in which the theme is shared out across several instruments—with modalities which in their negative aspects mirrored B's economy of expression and tendency to park aspects of her life in various disconnected annexes. I felt immensely sad, an affect which was so present within me and, apparently, so absent from her inner life. The piece was Webern's re-working of the theme of Bach's "The Art of Fugue"—fugue being the operative word, since it suggested a flight from something unbearable as well as infinite possibility.

I found myself rather shamefully retreating into wondering how the poet Akhmatova had survived and managed to be so creative during the time of her son's incarceration. Awakening from this reverie, I set to work trying to understand the fear, pain, or confusion that had driven B back into muteness. My offerings were ignored. The same happened on the second day, throwing me completely off balance from the tri-partite rhythm of re-entry, engagement, and parting, which I had found helpful in shuttle analysis. I also felt deeply concerned.

The next day, my interpreter arrived early to tell me that there was something I needed to know. Through the administrators of the project, B had made an offer of practical help to me, which I had refused, fearing a malevolent appearance of the Trickster. Her warm gesture of love and gratitude emerging from a thawing within herself and out of the analysis had been met by the only too familiar response from her childhood of an icy blast of mystifying rejection from me.

It took us some time to recover from this individual and collective attempt to dismember boundaries and create confusion in the borderlands between two analytic cultures. Without the information from our interpreter-cum-*whisperer*, I would have been left in the dark, identifying with both Akhmatova and her son, Lev—both imprisoned (one in her tiny flat on the Fontanka canal and the other in his cell), the two of them separated by a bridgeless river of grief, loss, and helplessness.

This pooling of countertransference, reverie, and the positive aspect of the *whisperer* helped us to recover slowly and came to form an aspect of work which I have found only in shuttle analysis. It suggested to me a "transcultural transcendent function" emerging from within the matrix of analysand, interpreter, analyst, and differing cultures. I conceived this deep connection arising in a multi-dimensional inter-psychic space, consisting of the following cross-over points:

(i) The Self-ego axes of analysand and analyst, which are personal and dovetail with the Collective-ego axes
(ii) The Collective-ego axes, which are transcultural and are built through archetypal imagery and affect

The Clash of Cultures

The depth and challenge of this sort of connection felt like a counterpoint to what Sebek has called "jolly analysis," which he understands as a reaction-formation against the totalitarian object.[5] This manifested itself in a small proportion of my work with analysands and supervisees as a flippant attitude towards the framework of analysis, its boundaries, and professional ethics. Lateness for sessions by analysands and puzzling extensions of time boundaries between them and their analysands (I did some supervision) beckoned constant attention but were sometimes tossed into the dump of bemusement, leaving me puzzled and frustrated.

I have no answer to my question, but wonder if the dissolution of the Soviet Union dismantled a previous sense of deep containment, albeit negative. Maybe "jolly analysis" was a signpost to the juxtaposition of an intellectual grasp of the horrors of the twentieth century alongside an awesome dissociation. Pertinent, perhaps, to these questions is this passage about dissociation from Jung:

> On the one hand, ego-consciousness makes convulsive efforts to shake off an invisible opponent (if it does not suspect its next-door neighbour of being the devil) while on the other hand it increasingly falls victim to the tyrannical will of an internal 'Government opposition' which displays all the characteristics of a daemonic subhuman and superman combined.[6]

Ethical Issues in Shuttle Analysis

"Jolly analysis" may also have been a way of coping with the framework of shuttle analysis with its constant yet rhythmic interruptions, some of which were experienced as repeated and sadistic abandonments. An ethical question made its entry at the point of my realisation that this particular form of therapeutic relationship was *potentially* harmful and could re-traumatise the very people into whose healing and individuation I had been invited.

Many of the analysands in the Russian Revival Project shared some, if not most of the features of this psychological photo-fit:

1. Trans-generational loss of the male line through war, exile, disappearance, and the siege (of Leningrad, now St. Petersburg)
2. Parental separation/divorce, often at an early age and leading to the loss of one or both parents

3. A close but often ambivalent relationship with a widowed *babushka* (grandmother) who commuted between human form and archetypal image
4. Several step/half siblings
5. Alcoholism in the family
6. Role modelling that upholds the inferiority of women
7. Masochistic surrender amongst women
8. Dependence cohabiting with chronic rebellion
9. Paralysis between creative initiative and heavy passivity
10. Problems with affect regulation within the context of insecure early attachment patterns

For analysands in whom a large number of these features appeared, sometimes with great emotional weight attached to them, I thought that the best way forward was to agree mutually to end the work and ensure that they found their ways to appropriate alternative and more supportive forms of therapy, which I did with the help of a psychologically aware General Practitioner (GP). In view of the absence of trained analysts, this seemed to me to be the ethical and appropriate therapeutic intervention.

The work had shown me that the long gaps between visits impeded the development of trust, dependency, and closeness, and could lead to very serious self/other-destructive behaviour such as abandonment of the self and dependents, or suicidal ideation. The privations of shuttle analysis tended to compound the analysands' physical and emotional deprivation and be too destabilising.

Mutual Assessment of Analysts and Candidates for Shuttle Analysis

Those best able to survive the demanding process of shuttle analysis had histories of positive, reasonably secure early and subsequent attachment patterns. Their embodiment was mirrored by a close relationship to their emotional, intellectual, and spiritual lives; and the combination of a symbolic attitude, ego strengths, and a developing Self-ego axis set the stage for exploring the horizons of the psyche, rather than having to revisit the personal and collective traumatic past.

Self-Assessment by the Analyst of Capacities and Limitations

The demands of shuttle analysis were not confined to analysands and my interpreter (whose work demanded very much of her). The process required of me physical, emotional, and intellectual stamina. Supporting that was Jung's notion: "*Ars totum requirit hominem!* ... A genuine participation, going right beyond professional routine is absolutely imperative, unless the doctor prefers to jeopardise the whole proceedings by evading his own problems, which are becoming more insistent."[6]

Always plumbing my conscious and unconscious motivation for undertaking shuttle analysis, I was aware of putting myself in guilt-inducing situations by disrupting and leaving my London analysands and doing the same with the people in St. Petersburg. The tri-partite rhythm of re-entry, engagement, and parting both held me and severely challenged me on both a personal and professional level and created problems for those at home and abroad, which have as their common denominator abandonment.

What I had not reckoned with was the impact of the amalgam of Russia's cultural complexes with my analysands' personal complexes and how those interdigitated with my personal history and inner world.

Conclusion

I have alighted very lightly on a few aspects of shuttle analysis in Russia from a subjective point of view. The process of shuttle analysis invites deeper and further exploration, as well as research, and this could be extended to other similar projects.

Does It Work?

If the closeness of the analytic relationship, albeit punctuated by long periods of absence, manages to create an inner loving, useful, and lasting presence of the analytic duo within both parties, my answer is *yes*. If it is in danger of re-traumatising the analysand in ways whereby the frame does not allow for adequate repair of rupture, then the answer is *no*. Our first duty is to do no harm.

THE "FRONTIER-LINE IN HUMAN CLOSENESS"
BY ANN FODEN

There is a frontier-line in human closeness / ... /
Those striving towards it are demented, and / If the
line seem close enough to broach— / Stricken with
sadness ... Now you understand / Why my heart
does not beat beneath your touch.

—Anna Akhmatova, *White Flock*

Not any religion or cultural system. / I am not from
the East or the West, / ... / My place is placeless, a
trace of the traceless. / ... / first, last, outer, inner,
only that / breath breathing human being.

—Rumi, "Only Breath"

INTRODUCTION

During the past twenty-five years programs of shuttle supervision and analysis have proliferated. In their attempts to spread and develop analytic understanding they are required to cross many frontier-lines—those of language, religion, Eastern and Western modes of consciousness, national prejudices and pre-conceptions, and cultural differences in experiences, history, and complexes. All of these warrant a detailed examination. At their core is what has been termed a "melody of separation" which, when it applies to shuttle analysis, carries a particular challenge to developed Western thinking about the necessity of creating a secure, reliable container in which both members of an analytic couple can feel safely held.[7]

To try and open oneself fully to this deep experience of *otherness* is primarily a shock and a disturbance to both parties, triggering both essential and also less helpful defenses. As is reflected above in the two brief extracts from longer poems, one of the core struggles for the shuttle analytic couple is to attempt to disentangle and bear the paradox of knowing that some of these frontier-lines can never be crossed whilst also recognizing that some aspects of the human psyche seem universal and familiar.

This means, however, that there can be no question about the mutuality of this experience. Shuttle analyst and analysand are truly thrown together from the outset into this threatening, potentially creative mix whether they realize it, initially, or not. In shuttle analysis there can be no doubt that the analyst is "a fellow participant in a process of individual development."[8] Susanna Wright, a shuttle analyst in St. Petersburg, summed this up: "My experience of threat from the unfamiliar was an important reminder for me of the vulnerability of my analysands to unconscious feelings of danger, outrage or shock when they engaged authentically in the analytic relationship."[9]

This chapter is based upon my experience of working with four analysands in Moscow for a period of over six years. Our face-to-face contact was brief, four separate periods of five days each year, during which they would have a double-session each day. Our periods of separation were much longer, anything from nine weeks to three months.

Whilst a rhythm of presence and absence is important in laying the foundation for a sense of self, the nature of this shuttle analysis rhythm impacted heavily upon both myself and my analysands, regardless of our differing analytic backgrounds. Whether we had experience of multi-session or single-session weekly work, all of us were used to regular, consistent contact. This shuttle analysis rhythm was more fractured and random with deep inter-subjective connection, severance, and reconnection all being dictated by an external timetable which, by definition, was unable to care for the inner life of the analytic couple. It could resonate all too easily with traumatic personal or collective experiences of attachment held deeply within either or both members of the analytic pair.

Given that face-to-face contact was so brief, and that external and internal experiences of space and time differ, the five-day periods of direct contact would feel more constricted than this. There was a "feast and famine" atmosphere to them with intense immersion and then departure. Some shuttle programs attempt to solve this dilemma by allowing for longer periods of face-to-face contact and whilst they may well produce less sharp edged starkness, they bring with them their own greater degree of disruption to the personal and professional lives of analyst and analysand. Shuttle analysis presented one all the time with the realization that the idea of continuous external analytic space and time is in itself a Western luxury.

In this brief chapter I will suggest some ways in which the psyche attempted to find a creative, purposeful way through these limitations.

Rage and the Capacity to Mourn

Visible through the window of the room in Moscow where I worked was Andrey Faydysh-Krandievsky's beautiful titanium Space Obelisk, erected to celebrate Yuri Gagarin's pioneering achievement. At 328 feet high, it depicts a huge plume of smoke propelling a tiny rocket away from the gravitational pull of the earth and into the sky. It always symbolized for me the extreme energy required for analysts and analysands to travel into this unknown shuttle analysis place and the fragility of the human container that had to survive these forces impacting upon it.

This chapter cannot do justice to the courage of shuttle analysands who have placed themselves within this capsule. Natalia Alexandrova,[10] writing about her own experience in struggling to trust that she could remain held within the reverie of the continually disappearing, shuttle analyst, cited Ogden's emphasis upon the importance of having time to waste.[11] To this I would add the importance of having time for negativity and time for silence.

Given that a sense of internal space is co-created by analyst and analysand, what helped us to try and achieve this? As an analyst it was my experience that there was an ever-present struggle to contain and process, not only the unfamiliar and the unknown, but also the guilt and ethical unease about the repeated abandonments and their potential for re-traumatization. It is right to note that the shuttle analysis "melody" is also one of repeated *returns* and these are undoubtedly important. But the idea that the frame provided the opportunity to revisit these traumas in the presence of a containing, processing other, who did not disappear forever, never felt fully convincing to either my patients or myself.

It was also very important to hold onto the conviction about the centrality of a boundaried, confidential *vas*, even when it felt as if it existed only as a metaphor within one's own mind. For my analysands two factors seemed especially important if they were to be able to benefit from this shuttle structure. One was the capacity to express outrage and protest at the very frame which held them, and the other was the capacity to symbolize absence and to mourn.

For those, like my Moscow analysands, who have experienced external and internalized totalitarian objects, an outward politeness and compliance becomes a survival necessity, matched by the equal psychic necessity of an inner, hidden defiance and opposition. But the deeper and more authentic the analytic connection, the more this split has to be relinquished.

One of my analysands raged at me about the shuttle experience:

> No food and then too much food. I don't understand you. Can't hear you. Don't feel anything. How can you? What do you think this frame does? You teach me I must survive during three months, then you come back. So what do you expect? We cope consciously, but emotionally you repeat trauma. Do you think a child can survive and grow like this with four weeks of parent there and no language?

Recognizing the guilt and shame that was activated in me she continued, "I make you sad. It is not enough that you feel grief—you have to be really hurt."

The profundity of those last words remained with me. Winnicott recognised the vital importance of rage and destructiveness being expressed and the analyst being able to absorb and survive it.[12] But my analysand was right, that, in this context, it was crucially important that I should allow myself to be "really hurt" at a depth not always visited or activated in my usual practice.

Repeated separations challenged my analysands' capacities to translate me from an absent external person into an internal representation that could be related to imaginatively. As Warren Colman has written, "the most fundamental requirement of mourning is the capacity to symbolize absence" and this, in turn, is dependent upon early experiences of containment which enable the pain to feel thinkable and bearable.[13] Where there is a history of traumatic loss or inadequate containment, absence can often be defended against only through dissociation or evacuation.

Russia is a country which carries within it a history of almost inconceivable loss. Joseph Brodsky, reflecting upon the dark years before the Second World War, wrote tellingly, "Death—became too real for any emotion to matter. From a figure of speech it became a figure that leaves you speechless."[14] All of my analysands carried within themselves personal and collective experiences of loss in varying degrees and my departures could trigger negative reactions such as dismissal, envious attack, and conscious refusal to contemplate my continued existence elsewhere. But these were essentially evidence of attachment, protective defenses used by those who were all too aware of what they were trying to ignore. This particular shuttle analysis frame could not assist where the past experience of loss had left an inheritance whereby the very idea of the importance of imagining a continued link with me seemed absent or meaningless.

Even here, however, it was my experience that the psyches of the shuttle analytic couple attempted to help in two distinct ways: by moving the work onto the level of enactment and also by moving it onto the level of the somatic. There were "moments of meeting" around objects, literature, or music, which felt deeply important and powerful.[15] To quote Colman again, "this might be regarded as interpretation through action, a creative form of enactment that transforms the relational field in a way that verbal interpretations alone cannot do."[16]

It was striking, also, how much of the significant interaction and pressure of shuttle analysis was registered in all of my analysands' bodies as well as my own. It was at times a very physically painful and uncomfortable place to be. But where there was a history of traumatic, unprocessed loss already lodged within the body, moments where I could bring my analytic focus onto the bodies of my analysands and register the impact within my own body, proved powerful and moving. With more vulnerable analysands at these crucial moments, where feeling and memory break through, the affect aroused can be very strong and the frame needs to be firm, stable, and consistent. I could never deliver this fully and, without the time and space to breathe, creative interaction could sometimes retreat into intellectual awareness, and the mutual somatic feelings could become suffocating and, at times, almost torturing. The bodies of analyst and analysand take a double hit when the pain of unprocessed loss meets the absences of shuttle experience.

Speech and Silence

"The act of speech in itself—can be seen—as an act of mourning for a lost connectedness."[17] Despite varying degrees of fluency in

English, all of my analysands chose the privacy of working without an interpreter, and so we communicated with the aid of a dictionary and a simplified, toned-down version of the language. In one sense it meant that none of us had access to the nuanced fluency and freedom of our mother-tongues.

Words could feel both a necessity and a hindrance. Often they felt the latter. Transference interpretations, although accurate on one level, could also feel misplaced, secondary, or irrelevant a lot of the time. Equally, amplification and association to dreams and images could feel, at times, rather sterile and manufactured.

It was here, in the pull away from words into sound, silence, and the somatic, that I felt most strongly the psyche's attempt to find a purposeful way through the shuttle experience. I noticed that my relationship with my own sound changed so that it moved onto different levels, becoming clearer, slower, and, at times, more emphasized. I was reminded of Anzieu's concept of the "sound mirror,"[18] the infant's earliest enveloping audiophonic skin, and Bisagni's description of the "rhythmic object ... crucial is the ability of the analyst to attune and deintegrate with the (autistic) analysand and to stay close to his or her rhythmical expressions."[19]

In the limited shuttle time available there is a heightened pressure on analyst and analysand to use words to establish meaning. But if this could be resisted and silence risked, there was often the sense of the work moving to a deeper level of connection and reverie.

By this I do not mean the silences of blankness and disconnection, but the hard work of coexisting within a wordless, processing, and embodied place. Sometimes these silences would feel psychotic and, often, full of helplessness or a form of suffocating deadness. Sometimes images would arise during these silences which held a frightening numinous power for an analysand and yet eluded conscious understanding. It was possible to interpret these images in terms of the deprivations of the transference and the shuttle frame, or aspects of the pre-verbal self, or personal and collective experiences. But however valid some of these may have been, the overriding feeling was that our words were rendering something lifeless and manufactured. It was like being held in a force field prohibiting interpretation. To hold onto this place meant resisting the urges of one's analytic superego to capture meaning and it also required holding at bay a sense of unease as to whether a trickster element was creeping in and sabotaging the work. If this uncomfortable, silent, *between* space could be sustained, then, often, the analysand's internal relationship to these images or symbols could shift, bypassing the need for conscious definition.

One of my analysands who experienced the pull of this place of cohabited silence paralleled it with Jung's story of the Chinese rainmaker.[20] She said, "Things are dry and disordered inside me. You come here from another country and we sit together all week in this empty room, and then things come back to order inside me and my rain comes." Later she said, "Sitting with you is a very authentic thing; you give me a wide space, a silence, an ocean; out of this comes the deep connections in my own work."

Rereading this story, what stood out for me was the *dried up* appearance of the rainmaker and the emphasis that is placed upon the necessity of him taking the disordered from outside into himself. Sometimes I have felt that the best I could offer to my analysands was this place of, seemingly dry, disordered *alongsidedness*. For me this links with what Clark was writing about, in the context of his work with regressed patients, when he described a consubstantial animating body "between us and inside each of us … [a] forceful psychoid level of experience," borne out of a "need to animate the other in order to animate the needy self, to make the other [mother/analyst] psychosomatically disordered in order to share, understand and so perhaps heal the patient's disordered self."[21]

Facing the reality of no shared language it was possible, sometimes, to find this broader place but it was costly and involved a loosening of one's familiar landmarks. Brodsky, again, wrote,

> One gets done in by one's own conceptual and analytic habits, e.g., using language to dissect experience and so robbing one's mind of the benefits of intuition. Because for all its beauty, a distinct concept always means a shrinkage of meaning, cutting off loose ends. While the loose ends are what matter most in the phenomenal world, for they interweave.[22]

Conclusion

This is a very small attempt to capture some aspects of a huge experience. Can shuttle analysis ever rise above or, perhaps more accurately, move below its inherent deprivations? Shuttle programs, with their inevitable compromises, are important and necessary. In true Jungian fashion, shuttle analysis stands poised between its proven capacity to do good and its potential to re-traumatize. As such, it requires continuing, careful assessment. I have attempted to show that a deep, caring connectedness can be achieved, but implicit in my illustrations are the requirements of an unsheltered openness in the analyst and a relatively stable ego functioning in the analysand. For some with deeper traumatic wounds and disturbances it may not be helpful as this requires a much greater continuity of contact and a shared language. But if the temptation to use cultural differences as a barrier can be resisted, and both analyst and analysand can risk losing the familiar ground under their feet, then, paradoxically, both have entrance into a broader, richer place, full of external and internal discoveries. I have been profoundly extended by this experience and I am grateful beyond measure to all who allowed themselves to enter into this place of authentic connection with me.

I will leave the last words with one of my analysands. Returning to Moscow after a week's sessions in London she had the following dream:

> *I dreamt that I bought a necklace in London for £40. It was a beautiful necklace with lots of jewels joined in a chain. But back in Russia I was told I had to pay three times this amount in tax—£120. I was very angry; it was not just. But I knew I was forced to because of the immigration laws and, if I did not, I would never be allowed a visa to come back to London. There was no choice.*

For the privilege of reaching out across so many frontier lines, shuttle analysis does indeed seem to demand of both analyst and analysand that they are willing to attempt to pay three times the price.

NOTES

1. Catherine Crowther and Jan Wiener, "Finding the Space between East and West: The Emotional Impact of Teaching in St. Petersburg," *Journal of Analytical Psychology* 47 (2, 2002): 285–300.

2. C. G. Jung, "The Complications of American Psychology," in *The Collected Works of C. G. Jung*, vol. 10, ed. and trans. Gerhard Adler

and R. F. C. Hull (Princeton, NJ: Princeton University Press, 1930), § 972.

3. Orlando Figes, *The Whisperers: Private Life in Stalin's Russia* (London: Penguin Books, 2007).

4. Igor. M. Kadyrov, "Analytic Space and Work in Russia," *International Journal of Psychoanalysis* 86 (2, 2005): 467–82.

5. Michael Sebek, "Psychoanalytic Training in Eastern Europe," *Journal of American Psychoanalytic Association* 47 (3, 1999): 983–88.

6. C. G. Jung, "The Psychology of the Transference," in *The Collected Works of C. G. Jung*, vol. 16, ed. and trans. Gerhard Adler and R. F. C. Hull (Princeton, NJ: Princeton University Press, 1946), §§ 394, 400.

7. Gábor Szönyi and Tamara Štajner-Popović, "Shuttle Analysis, Shuttle Supervision, and Shuttle Life—Some Facts, Experiences and Questions," *Psychoanalytic Inquiry* 28 (3, 2008): 328.

8. C. G. Jung, "Principles of Practical Psychotherapy" (1935), CW 16, § 7.

9. Susanna Wright, "In the Missionary Position: A Perspective on an Alien Self," (lecture, Borderlands, 2nd IAAP European Conference, St. Petersburg, 2012).

10. Natalia L. Alexandrova, "The Double Burden of Shuttle Analysis," *The Journal of Analytical Psychology* 56 (5, 2011): 644.

11. Thomas H. Ogden, "On Psychoanalytic Supervision," *International Journal of Psychoanalysis* 86 (5, 2005): 1271.

12. Donald W. Winnicott, "The Use of an Object," *International Journal of Psychoanalysis* 50 (4, 1969): 713.

13. Warren Colman, "Mourning and the Symbolic Process," *The Journal of Analytical Psychology* 55 (2, 2010): 29.

14. Joseph Brodsky, "The Keening Muse," in *Less Than One: Selected Essays* (London: Penguin Modern Classics, 2011), p. 49.

15. Daniel Stern et al., "Non-interpretive Mechanisms in Psychoanalytic Therapy: The 'Something More' than Interpretation," *International Journal of Psychoanalysis* 79 (5, 1998): 913.

16. Julia Borossa, "Languages of Loss, Languages of Connectedness," in Judit Szekacs-Weisz and Ivan Ward, eds., *Lost Childhood and the Language of Exile* (London: Imago MLPC and The Freud Museum, 2004), p. 30.

17. Colman, "Mourning and the Symbolic Process," p. 284.

18. Didier Anzieu, "The Sound Image of the Self," *The International Review of Psychoanalysis* 79 (6, 1979): 23.

19. Franscesco Bisagni, "Out of Nothingness: Rhythm and the Making of Words," *The Journal of Analytical Psychology* 55 (2, 2010): 257.

20. C. G. Jung, "Adam & Eve," in *The Collected Works of C. G. Jung*, vol. 14, ed. and trans. Gerhard Adler and R. F. C. Hull (London: Routledge and Kegan Paul, 1966), § 604.

21. Giles Clark, "The Animating Body: Psychoid Substance as a Mutual Experience of Psychosomatic Disorder," *The Journal of Analytical Psychology* 41 (3, 1996): 367.

22. Brodsky, "Less than One," p. 31.

CHAPTER 13

AN EAST-WEST *CONIUNCTIO*
THE RELATIONAL FIELD IN CROSS-CULTURAL ANALYSIS

LIZA J. RAVITZ

Coniunctio—the alchemical term used to describe the union of different substances.

When I was a small child, growing up in New York, Sundays were special days when my family would go to the beach and I would dig in the sand. My father would say, "If you keep digging perhaps you will reach China. It is very far away." I understood this to mean that China was on the other side of the world, an unknown and exotic place. After the beach, we would go to our favorite Chinese restaurant for dinner; in my family, Chinese food was our favorite food. When my mother and I would spend our special time together, she would playfully say she believed she was a Chinese princess in another life. These early experiences brought China into my consciousness at a very young age. I understood China as familiar, for I had eaten her food, yet it was at the same time, very far away, very exotic, mysterious, and *Other* to me.

In college during the 1960s, I began my love affair with Far Eastern religions and began to meditate. Through the influence of cultural icons of the times, interest in Eastern philosophy flourished on American college campuses and became an influence in the development of the *Hippie culture*. Through graduate school and my early psychologist career I continued my meditation practice and interest in Buddhist philosophy but never connected either to psychology. I was plagued by a seemingly irresolvable split in my career interests between psychology and religious studies. My first true meeting with Jung came through my interest in a Taoist alchemical text, *The Secret of the Golden Flower*.[1] When I read Jung's commentary on the connection between his psychology and Eastern philosophy, I had an epiphany, one of the most important experiences in my life. Through this experience, an avenue for the resolution of my inner conflict was revealed. Jungian psychology incorporated both interests, a *coniunctio* had occurred in my psyche as the opposites were bridged. This experience set me on my path to becoming a Jungian analyst. For me, Jung and the East were then forever connected.

As synchronicity and fate would have it, thirty years later, from 2012–2014, I found myself living in the Far East, in Taiwan, practicing Jungian analysis in the capacity of visiting analyst for the Jungian Developing Group of Taiwan, under the auspices of the International Association for Analytical Psychology (IAAP). My analysands were Chinese/Taiwanese adults, males and females, between the ages of thirty-six and sixty-five. All spoke English as a second language. They were seen at a variety of frequencies, mostly twice a week. Out of the alchemical mix that took place in the interactional field between us, and in spite of our differences, a transformative process took place, generating meaning and the experience of greater wholeness. How was it possible to develop such intimate, healing relationships when the differences in our cultures seemed so daunting? This chapter will cover my reflections on this question.

When analyst and analysand come together in a relationship, there are always differences and complications to negotiate. When added to this there is the further complication of diverse cultural, ethnic, and language differences, the practice of analysis becomes even more complex for it requires two people to come into relationship with each other in a very intimate way. Communication involves a verbal exchange of an extremely nuanced character as well as the mutual reading of body language. The psychoanalyst relies upon and reflects on the verbal and nonverbal information to more fully understand the analysand. Shared culture is a bond which contributes to comfort and familiarity in the relationship. In America, for example, mention of George Washington, Thanksgiving, or 9/11 are common signifiers that have collective meaning to both people. Humor is often understood through *double entendre*, local dialect, or trendy slang. In this situation, there are some basic similarities in the cultural unconscious of both participants which enter the work, laying both a conscious and unconscious rudimentary foundation for mutuality. But what happens when the analyst and patient come from very different cultures and speak different languages? One party must use a second language in order to communicate while the other communicates in their native tongue or both must find a mutual second language. What are the consequences to mutual understanding when the common signifiers that a common culture and language bring are missing? In many ways, in this situation when the diversity is so obvious, the two people involved meet as true strangers, with perhaps no common overlap except their humanity and interest in psychoanalysis.

In Jungian theory, the transcendent function is generated by the tension of opposites being held and endured until deeper layers of the psyche respond.[2] In my analytic work a variety of dichotomies emerged; this took the form of Chinese culture vs. American culture, Eastern mind vs. Western mind, and the collective unconscious vs. the personal unconscious. The language differences generated many opposites such as indirect communication vs. direct communication, implicit communication vs. explicit communication, and understanding vs. not understanding. The opposition of the archetypes of the *Universal* versus the *Other* was omnipresent. All of these tensions entered into my experience of practicing Jungian analysis in the Far East. What meaning can I find for this constellation of opposites?

William Blake, the renowned English artist and visionary, suggests contraries are a necessary ingredient for human existence. He writes in the *Marriage of Heaven and Hell*, "Without contraries there is no progression. Attraction and repulsion, reason and energy, love and hate, are necessary to human existence."[3] Carl Jung, in a similar manner, stated that the opposites "are the ineradicable and indispensable preconditions of all psychic life."[4] Jung chronicles, in *The Red Book*, his process of uniting ego consciousness with the unconscious.[5] The result of this union was his discovery of the self, the archetype of wholeness. Jung concluded that the very essence of a human being is made up of the struggle to negotiate the opposites. It is through the tensions and unions of these opposites that the generative capacity of the human being emerges, and it is through this process that he or she discovers the deeper self and makes meaning of his life.

The Western analyst practicing in an Eastern culture, by definition, is *Other*, a stranger, a foreigner, the opposite. This *stranger* is perceived as foreign because he comes from a different land, speaks a different language, and is from a different culture. Yet he introduces new ways of thinking and perceiving for "he looks at us with a different eye, a new eye, and tells us what we no longer see about ourselves; therefore he helps us rediscover ourselves."[6] This activation of the *Other* archetype entered the analytic relational field immediately fostering the creation of an atmosphere that was out of time and space. There seemed to be an understanding by my analysands that I, as a foreigner, was here for the moment but would eventually leave, go back to another land, possibly never to be seen again. As I was from another culture, I did not follow Chinese social status rules. This had the effect of freeing my patients from many of their cultural restrictions of interaction. In my experience, this promoted the lowering of defences and created a unique space for the surfacing of long repressed shadow material. Barriers that might have impeded analytic work, reduced by our cultural statuses, allowed the processing of material that might have taken years to access had the pair been culturally matched. Being with a foreigner is a little like wearing a mask in that it gives permission for other parts of the self to come forth. With a foreigner, there seems to be an unconscious pull to uncover what heretofore has been taboo to express; prohibitions reinforced by culture are loosened, the adherence to cultural restrictions begins to break down, and fear of ostracism and reprisal for not conforming to cultural norms diminishes. The results of diversity that are intrinsic to a mixed analytic pair invite this kind of flexibility. The reduced barriers and the permission given by the unfamiliar open up the analytic space. This open space allows fluidity to the interaction, which in turn encourages the play of imagination, activating the urge toward individuation.

On the other hand, if difficulties in understanding developed, sometimes my analysands would opine that the reason was our different cultures. At times this may have been true, but I felt sometimes invoking our different cultures was a form of resistance, a way around having to directly wrestle with our miscommunications. Teasing out the sources of our misunderstandings was challenging without the capacity to use nuanced language, coupled with the Chinese cultural tendency to avoid conflict and create harmony. In this situation, I had to rely on my intuitive and feeling functions to make a determination. Along with my patients, I also experienced a larger space that invited the movement of imagination. The loosening of restrictions invited me to flexibility in analytic method and to the development of new ways of relating, communicating, and understanding. The challenges of language and custom prompted me to examine my beliefs and method in regards to practicing analysis and to create for myself new verbal and non-verbal ways of understanding and communicating. In order to do this I had to further develop my intuitive function and fine tune my ability to generate metaphor in more nuanced, far-reaching ways. Many times it was only through these channels that I could understand what was unfolding in a session, because often the nuances of verbal communication were missing. I will give several examples below.

In understanding and making meaning of the stories of my patients, there was always an omnipresent cultural pair of opposites operating between us: that of the Eastern *collective* perspective versus the Western *individual* perspective. Here is where the Western and Eastern experiences are most diverse. The Chinese cultural tradition places supreme value on harmony in the collective. The smooth functioning of the group, be it nation, community, or family, takes priority over individual needs. Individual expression is suppressed for the needs of the group. The Western cultural tradition in America in particular, has a more individualistic value. Individual needs, goals, and achievements are emphasized. Compromising individual strivings for the good of the collective is not as prominent a value. These differing perspectives are woven into the very fabric of social, political, and psychological life in each culture.

We see these differences operating in the cultural unconscious of both cultures as, for instance, we examine each culture's *Dream*. The American Dream emphasizes the individual. In his book *Epic of America*, James Truslow Adams coined the term.[7] He defined the American Dream as "a dream of social order in which each man and each woman shall be able to attain to the fullest stature of which they are innately capable, and be recognized by others for what they are, regardless of the fortuitous circumstances of birth or position."[8] The Chinese Dream, on the other hand, emphasizes the collective. The Chinese Dream according to China's President Xi is about "realizing a prosperous and strong country, the rejuvenation of the nation and the well-being of the people."[9] According to the state-run news agency Xinhua, it means that all workers should "combine their personal dreams ... with the national dream and fulfill their obligations to the country."[10]

This difference in perspective is imbued in the social structure of each culture. Walking recently on an American campus, I encountered a huge banner shouting, "STAND OUT AND BE COUNTED." This slogan would be foreign to the Chinese culture where humbleness, cooperativeness, and harmony in the group are prized over individuality.

This dichotomy was often constellated in my analytic work with my Chinese/Taiwanese patients. The essence of Jungian analysis is individuation, finding one's individual path, separate from the constraints, expectations, and dictates of the collective. A Western notion of individuation is in direct opposition to the traditional values of the Chinese culture. As we shall see, traversing these waters of opposition, at times, created conflict and confusion in my patients.

China/Taiwan is in a difficult time of social change now as the influences of the Western world are increasingly felt. Recently, when Chinese individuals were asked about the Chinese Dream, a similar vision to that of the American Dream emerged, that is, interest in enhancing the well-being of the individual.[11] Stirrings of an urge toward individuation, towards more emphasis on the individual are beginning to be felt. As Asia becomes more westernized, the integration of these conflicting values becomes a major psychological challenge.

In the social sphere, these differences in perspective manifest themselves in differing Western and Eastern family practices which have implications for developmental theory. The parental complexes in Chinese culture evolve from experiences that are very different from those in America. For example, traditionally, in the Chinese culture, children sleep in the same room with their parents and continue to live with their parents after reaching adulthood. A young man will bring home his bride to live with his parents. Home life tends to be communal, including extended family who often live in the same house. The traditional Chinese culture places great emphasis on honoring, respecting, and listening to parents. Parents have deep and integral influence over the choices their adult children make about marriage, career, and child-raising. This differs from Western culture, particularly America, where emphasis is placed on a growing independence from parents as the child becomes more adult. For instance, the Western developmental norm places a high value on children eventually sleeping in their own beds in their own or sibling-shared rooms. Launching as a young adult means physically leaving the parental home, choosing a career, and finding a mate of one's choice. The expectation is that a healthy young adult, as he separates from his family, will begin to make his own decisions about the future of his life. Recently, much press has been given to the fact that young adults are not leaving home but, because of economics, continuing to live at home. This deviation from the pre-existing norm has been marked as a prolonged adolescence and has been seen as a regression in development.[12] According to Western developmental norms, Chinese family life would be seen as regressive and unhealthy. When practicing cross-cultural analysis, this bias in cultural norms can have a negative impact on how the analyst thinks about her analysand's development. It is incumbent on the analyst to become aware of cultural differences and adjust theoretical thinking about normative behavior accordingly.

In working with Chinese/Taiwanese patients, a psychological conflict often arose over this collective versus individualistic point of view raising many questions for them. Do I sacrifice my needs for the good of my family and accept and conform to traditional values or do I follow my own path, make my own decisions for the future, and separate from the needs and wants of my parents and family? Do I choose a career I am interested in or one that is compatible with the needs of my family or the social status of the culture? Do I raise my child as I wish or follow my parents' dictates? In recent years, with broader exposure to Western values, this inner conflict has deepened and become more prevalent. To break with these collective values is anti-cultural, not just an individual struggle as in the West. In Chinese culture, Jungian analysis, with its emphasis on individuation and finding one's own path, is an anti-establishment endeavour. Many conflicts emerge—conflicts that are not solely based on personal psychology.

How does a patient deal with these conflicting values and manage to hold and integrate the opposites?

For example, an unmarried, forty-year-old, female analysand who grew up in a very traditional family system was accepted to a graduate program. From her point of view, the expectation of the culture was for a woman to get married, have children, and take care of her in-laws. If the woman did not marry, she was expected to remain at home and take care of the extended family living in the house. By having individual aspirations and showing her academic abilities, this analysand felt she was showing too much pride. The culture does not support a woman showing her intellectual abilities. Her relatives asked why she needed so much education. They opine it will not be easy to marry. "Sometimes I feel angry and sometimes I feel narcissistic and not humble, deficient in following collective values, and obeying my parents. In my family, I am the first woman to not obey the rules for women about marriage, children, and education." An outer solution was reached to study abroad in a Western country, thus bypassing cultural pressures; however, resolving this ongoing dilemma at an inner level requires the holding of the tension of the opposites until a resolution reveals itself.

The Chinese/Taiwanese culture has many collective rules about how to behave with others. There are certain expectations for relationships between parents and children, between teachers and students, between older and younger generations, and between men and women. These behaviors are habitual and automatic. As a Westerner, I am aware that my behavior and verbal responses did not necessarily follow these unspoken collective rules and thus resulted in the forming of a different kind of relationship than if I had been of the culture. For example, if I were Chinese, the relationship with my analysands would have been more formal. As the analyst, I would never be called by my first name. Sometimes people bow to each other in greeting, the angle depending on social status. I did not bow. Expression of negative feelings about parents, disagreement with an elder, or talk of sex and anger are strongly discouraged in the culture. I was told by a few analysands that they would never discuss sex with a Chinese therapist. I encouraged expression of these topics. Through opening up these heretofore taboo subjects, tacit permission was given for experimentation with new ways of thinking and behaving. My analysands began to evaluate their values and to find their authentic selves within the persona of the collective.

For instance, one analysand, a forty-two-year-old, single female, began to question her relationship with her father who was extremely controlling and verbally abusive. She felt she could not breathe when he was in the house, that there was no space for her to exist. It had never entered her mind that she could have an impact over this situation; in the Chinese power hierarchy, a single adult woman has little power in the family. At first she was hesitant to talk about her negative feelings towards her father as she felt this broke a collective taboo. Through ongoing discussion of her situation, a larger internal space for reflection was created inside her. This helped her to claim more space inside her father's house and an eventual resolution of moving into her own house. During this process, she had to decide which social and familial rules she could replace with her own individual values.

This same process has gone on inside me as well, resulting in much reflection on my values and behaviors for correct living in society. At times, I have had to adopt a more collective perspective while listening to my analysands' stories. This expanded view helped me to be more flexible in my thinking and to understand life situations better. The depth and breadth of this cultural difference has to be digested and embodied by the Western analyst. The Western analyst must wrestle with this cultural difference and develop within her own psyche a broader attitude of sacrificing individual needs for the good of the collective. As the analyst reaches an inner *coniunctio*, the opportunity for a larger mutual psychic space is created. This supports the process of the Eastern patient to bridge conflicting perspectives and transform suffering into something that creates meaning out of his experience. This is an ongoing mutual process that operates in the relational field between us as we co-create rules for our relationships.

Although the differences were visible and out in front of consciousness, there were also similarities to be discovered. To my surprise the dominant feeling of doing analysis with my Chinese/Taiwanese analysands was the discovery of commonalities and universals.

We very quickly deepened into the terrain of the collective unconscious.

Figure 1. Illustration from the sixteenth-century alchemical treatise *Rosarium philosophorum*.

In spite of language differences, I felt a deep empathic resonance with my patients. The image (Figure 1) from the *Rosarium philosophorum* of the King and Queen joining their left hands speaks to this unconscious energetic exchange.[13] In *The Psychology of the Transference*, Jung writes in regards to this image: "The contract of left hands could ... be taken as an indication of the affective nature of the relationship."[14]

There is universality in human suffering that comes with different phases of development. The dependency of childhood, the separation issues of young adulthood, the search for meaning in midlife, the vicissitudes of aging, and the spiritual preparation for death all have universal aspects regardless of culture. These archetypal themes appeared in my analysands' conversations, dreams, active imaginations, and sandplays. Through these life narratives, we began almost immediately to travel the waters of trauma, early childhood trauma, intergenerational trauma, and cultural trauma. These types of trauma are universal in nature and live in the collective of human suffering. Issues of attachment and the early maternal environment were common themes. These early experiences, both positive and negative, affect the developing psyche regardless of culture.

The cultural trauma suffered by the Chinese and Taiwanese is quite profound. Oppression, persecution, and flight from the homeland, with a resulting dispersal of Chinese people around the globe, have been a common theme in the histories of my Chinese/Taiwanese patients. Most recent memories include the tumultuous times in Taiwan during the transition to the Chinese government of Chiang Kai-shek and the Cultural Revolution in China. In December 1949, Chiang Kai-shek, in conflict with Mao Tse-tung, lost his political hold in China and moved his government to Taiwan. This period, from 1949 until 1987, was known as the "White Terror" during which many citizens in Taiwan, especially the intellectual elite, were tortured, arrested, and executed. In China, during the Cultural Revolution of 1966–1976, led by Mao Tse-tung, millions of people were persecuted, tortured, publicly humiliated, and displaced from their homes. At a glance, this might seem to be an experience a Western analyst could never fully understand; however, in many ways, the American experience has similarities. America was built largely by people who fled their home country because of persecution and oppression. This phenomenon continues today. Almost everyone in the United States has some story in their family history of heroic escapes from persecution in their homeland to a new life in America. It is part of the American cultural unconscious.

Coming from a Jewish Eastern European heritage, this theme is quite familiar. The Jewish cultural experience also includes traumas around oppression, persecution, brutality, and flight from one's home. In my cultural background, the continuous persecution of the Jewish people throughout history from before the birth of Christ to the present resonates with the persecution of the Chinese throughout the ages. The Chinese, like the Jews, have dispersed all over the globe. Both groups have had to adapt to foreign cultures while maintaining their own cultural heritage. In practicing analysis with Chinese/Taiwanese patients, I find that these cultural experiences resonate in their personal histories, impacting on the development of the psyche. As they tell their stories, a resonance takes place in my psyche too. There is a meeting of experience, a *coniunctio* that enhances a connection to the self for both of us.

There are further similarities in the very basics of the cultures. Both cultures mark the passing of time on the lunar calendar. Many of the Jewish and Chinese festivals fall on the same day, connected to the phases of the moon, inherently celebrating the cycles of nature. This lunar influence lives in the unconscious of both cultures. In terms of language, both Hebrew and Chinese characters carry symbolic as well as linguistic characteristics. All these similarities create some degree of overlap in the cultural unconscious of my patients and of myself, as well as carrying archetypes that live in the collective unconscious universal to us all. These factors help form a common ground where our unconscious communications can meet and unite in the container of the interactional field.

I would now like to turn to the issue of language. One might imagine that language can be the greatest obstacle to a successful analysis when a native English-speaking analyst works with an analysand for whom English is a second language. My experience has been that even when both partners make efforts toward mutual understanding, there are times when we cannot fully explain our experiences and grasp at words that may not fully express accurate meaning. At that moment a gap in communication exists. Here is where indirect and non-verbal communication takes a more prominent place. This diagram (Figure 2) speaks to what happens in a deep analysis.

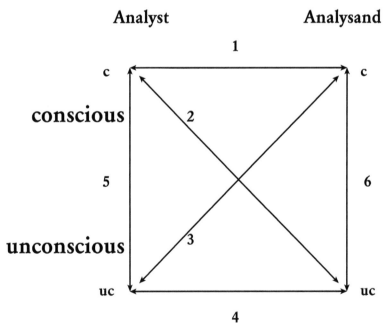

Figure 2. Lines of communication. [Adapted from Jung's "The Psychology of the Transference," CW 16, § 422]

In this schema there are six communication possibilities going on in an analytic session. Four are between analyst and patient: 1) conscious to conscious; 2) analyst conscious to patient unconscious; 3) patient conscious to analyst unconscious; and 4) unconscious to unconscious. Two communications are within the individual: 1) analyst conscious to analyst unconscious; and 2) patient conscious to patient unconscious. Of these six communication possibilities only one is conscious communication to conscious communication. The remaining five all include some aspect of unconscious communication. This type of communication does not include words but involves an energetic exchange *between* patient and analyst, as well as *within* the individual psyches of both. This energetic exchange is a universal language understood by all cultures. When language is more of an issue in analysis, more reliance falls on the non-verbal communications for understanding and healing. The analyst needs to cultivate an attitude of receptivity and openness to the unconscious of both the patient and herself. Here countertransference communications are especially important clues to the messages being transmitted by the patient. These communications can be seen as analogous to the communications between mother and infant.

In the last few decades there has been a convergence of agreement through research and study about the human mother-infant pair. This convergence comes from different disciplines including analytic infant observation, developmental psychology, attachment theory, and neuropsychoanalysis.[16] The agreement centers around the impact of the earliest mother-infant relationship on the developing self of the infant. We now know from brain research that this earliest of relationships sets down neural pathways in the brain that become the prototype for all future relationships, affecting how and if we will attach and bond with another. Through infant observation studies it has become clear that the self is not fully formed at birth but emerges over time in contact with its environment. Thus, development of the self is dependent on the earliest maternal relationship and becomes hardwired in the brain.[17]

In the healthy mother-infant pair, the baby, through the unconscious process of projective identification, alerts the mother to its inner state.[18] The mother, receiving this information, contains it in her mind, digests it, and takes action transforming the projected state into something more tolerable for the baby. Through this right-brain to right-brain unconscious communication, the mother is able to care for her infant and regulate his emotions thus establishing a feeling of safety and security in the baby. Through this process psychic space is created in the infant for him to eventually regulate his own emotional states. This process sets the stage for all future relationship patterns. Thus, for a healthy self to emerge, there is a need for emotional receptivity from the mother.

We can liken this process of the mother-infant attunement to the relationship between patient and analyst. It is the attuning, witnessing, reflecting analyst who, in an analogous manner, contains and mediates her patient's intense emotional states which are communicated to her through the unconscious phenomenon of projective identification. As the analyst digests and holds these infantile states within her own mind and body, a feeling of safety, protection, and regulation begins to develop within the patient. It is through this attunement that the patient begins to experience something different from the original deficient mother-infant communication pattern. With this healthier communication, repair of the damaged self can begin to take place and a healthier attachment pattern can be established. This allows for deeper connection to the unconscious by the patient and opens up the possibility of a free flow of psychic energy between the ego and the self.

When analysis reached a verbal communication gap, reliance on the non-verbal became necessary. For example, the use of metaphor was helpful for mutual understanding. On one occasion to help with communication I drew upon the English proverb, "The squeaky wheel gets the oil." In response, I learned that there is an equivalent Chinese proverb: "The noisy child gets the candy." Or for another situation when an analysand felt isolated as the boss at work, I said, "It's lonely at the top," to which she told me there was a line in a Chinese poem: "At the heights it is cold." Here there was a meeting of East and West, a union of opposites. The essence of these conversations was understood by both of us.

Images also supported bridging our mutual understanding. When a woman began to have a change in attitude from other-centered to inner-centered, in order to explain her experience, she invoked the image of the New King (Figure 3), saying "My clothes are still too big but I am emerging."[19]

Figure 3. Plate VII of the alchemical treatise *Splendor Solis*.

Sandplay, another way of engaging analysands around images, is popular in Asia and has a following in Taiwan. I found my Taiwanese analysands to be drawn to symbols and images and were comfortable using sandplay. As an introverted, non-verbal approach, sandplay suited the Eastern psyche. As a non-verbal means of expression, it invited psychic return to earlier, preverbal times. It is at this layer that the sandplay analyst can hold the maternal space. When this relationship becomes internalized by the patient, a basic maternal matrix is created or augmented upon which the total personality can develop.[20] At times, when an analysand brought a dream, the English translation was quite simplified losing its full expression. But often the dream would lead to an active imagination, an image, or a spontaneous association which opened a deeper connection to the unconscious, moving the individuation process forward. An example of this occurred one afternoon. When we began the session she seemed very nervous. It was taking us time to settle in together. She told me a dream without much emotion.

> *I am in class, about to take an exam. All of a sudden I see the night lights of the city. Then they turn into stars in the sky that sparkle with a golden light. I reach for the gold stars but as hard as I try I cannot grab them. My spirit knows they contain wisdom but I cannot reach. I sit before the exam, frozen.*

On hearing the dream, my inner thoughts associated to gold and alchemy. I was curious about the gold stars. We explored the dream a bit but there was no spark of connection between us. I then asked her to draw her dream for me. As she started drawing, I felt a gush of sadness enter my body. Finally! As she was showing me her picture, I told her I felt some sadness about the dream. In response to this she told me the gold stars were knowledge that she could not quite grasp but knew was there. She brought up her deep sadness about underachieving in school. Because of this, she had earned the disappointment of her father. This conversation opened up a whole new frontier for us of looking at her father complex.

Whether we drew upon metaphor or image, my experience gave deeper meaning to Jung's statement, "An archetypal content expresses itself, first and foremost in metaphors."[21] I have found that even with the limitations of language, this process of contacting the self, in the context of the therapeutic interactional field of the patient and analyst, takes place in ever deepening ways. Our communication was "good enough" for deep transformative processes to take place.

Recently, when a Taiwanese person referred to me as ethnically American but spiritually Chinese, I felt a resonance inside myself. When I returned to America after a year in Taiwan, many of my colleagues commented upon seeing me, that I looked Chinese. One person said, "Your face has changed, you even walk Chinese." Another person said, "Your face is calm, very Chinese." Whatever these statements really meant, even as projections from my colleagues, there must have been some energetic exchange between myself and my Chinese/Taiwanese analysands for people to comment in this way. To my mind I have experienced a deep transformative process. It is difficult to articulate this exactly as I am still digesting and integrating my experience but I find that I am more authentic, more patient, more present, and more intuitive in my analytic work. Once again, I have experienced a true East-West *coniunctio*. I have dug down to China.

Cross-cultural analysis presents many challenges; however, if the analyst can become aware of nuanced cultural differences, develop her intuitive function, and pay attention to countertransference reactions, transformative processes within both patient and analyst can take place. For myself, I have been forever changed; I have lived in Taiwan, and, now, Taiwan lives in me. "When new bloods mingle, there appears a new space, a new language, a new culture and a new knowledge bridging yesterday and tomorrow."[22]

NOTES

1. Richard Wilhelm, trans., *Secret of the Golden Flower: A Chinese Book of Life* (New York: Harcourt Brace Jovanovich, 1962).

2. C. G. Jung, "Psychological Types," in *The Collected Works of C. G. Jung*, vol. 6, ed. and trans. Gerhard Adler and R. F. C. Hull (Princeton, NJ: Princeton University Press, 1971), § 827.

3. William Blake, *The Marriage of Heaven and Hell: A Facsimile in Full Color* (New York: Dover Publications, 1994), p. 1.

4. C. G. Jung, *Mysterium Coniunctionis*, vol. 14, *The Collected Works of C. G. Jung*, ed. and trans. Gerhard Adler and R. F. C. Hull (Princeton, NJ: Princeton University Press, 1963), § 206.

5. C. G. Jung, *The Red Book: Liber Novus*, trans. M. Kyburz, J. Peck, and S. Shamdasani (New York: W. W. Norton & Co., 2009).

6. Nouk Bassomb, "The God with Two Faces," *Parabola* 20 (2, 1995): 66.

7. James Truslow Adams, *The Epic of America* (New York: Little Brown, 1931).

8. *Ibid.*, pp. 214–15.

9. Clarissa Sebag-Montefiore, "The Chinese Dream," *New York Times*, May 3, 2013, Latitude.

10. *Ibid.*

11. Yan Sophia, "The Chinese Dream: It's Surprisingly Familiar," *CNN Money*, Jun. 10, 2014. Accessed Sept. 15, 2014, at http://www.money.cnn.com/2014/06/10/news/economy/chinese-dream.

12. Adam Davidson, "It's Official: The Boomerang Kids Won't Leave," *New York Times*, Jun. 20, 2014, Magazine.

13. Joseph Henderson and Diane Sherwood, *Transformation of the Psyche: The Symbolic Alchemy of the Splendor Solis* (New York: Brunner-Rutledge, 2003), p. 50.

14. C. G. Jung, "The Psychology of the Transference" (1946), in *The Collected Works of C. G. Jung*, vol. 16, ed. and trans. Gerhard Adler and R. F. C. Hull (London: Routledge and Kegan Paul, 1966), § 49.

15. Mario Jacoby, *The Analytic Encounter: Transference and Human Relationship* (Toronto: Inner City Books, 1984).

16. Jean Knox, *Self-Agency in Psychotherapy: Attachment, Autonomy, and Intimacy* (New York: W. W. Norton & Co., 2010); Louis Cozolino, *The Neuroscience of Psychotherapy: Healing the Social Brain* (New York: W. W. Norton & Co., 2010); and Allan Schore, *The Science of the Art of Psychotherapy* (New York: W. W. Norton & Co., 2012).

17. Knox, *Self-Agency in Psychotherapy*.

18. Schore, *The Science of the Art of Psychotherapy*.

19. Henderson and Sherwood, *Transformation of the Psyche*, p. 74.

20. Knox, *Self-Agency in Psychotherapy*.

21. C. G. Jung, "The Archetypes and the Collective Unconscious," in *The Collected Works of C. G. Jung*, vol. 9i, ed. and trans. Gerhard Adler and R. F. C. Hull (New York: Pantheon Books, 1959), § 267.

22. Bassomb, "The God with Two Faces," p. 67.

CHAPTER 14

GIVING VOICE TO PSYCHIC PAIN
THE BRITISH-MEXICAN CONNECTION, ON THE VICISSITUDES OF CREATING A HOME FOR STREET CHILDREN

Alessandra Cavalli

Introduction

In psychodynamic thinking the concept of containment is a basic one. It starts in infancy when the mother contains her baby in her womb, and allows, together with her husband, a place in their minds to develop so that their infant can be thought about. It is from their minds that father and mother monitor their infant's physical and psychological growth. It is thanks to this experience that the infant will discover that he too can develop the awareness of being a person with a mind from which he can contain his feelings, his emotions, and his thoughts.[1]

There are many children who have never been able to develop a mind as such; indeed, to have parents able to bring up their children with their psychological needs in mind is a great privilege. Unfortunately, there are children for whom physical well-being is already a luxury. For these children survival is a priority as there is no adult there to care for them physically, let alone emotionally.

There are countries in which the State overcomes this kind of problem by setting up social interventions to offer children protection and to compensate for what parents might fail to provide, but there are other countries in which the State is socially and economically not organised to offer an alternative solution; issues like "protection" of the child's survival are left to private organisations such as the Church, Non-Governmental Organisations (NGOs), or single individuals. Although this chapter will deal with the practical effects of this problem, a discussion of its implications is beyond the scope of this work. In particular, I will describe my work in a Mexican city and how an NGO working with street children has been able to move from offering protection and ensuring survival to the children they are working with, to helping children to develop minds of their own. In this, they try to provide something that for the normal development of children is expected.

From Protection to Containment

In the 1970s a Brazilian Catholic priest, Paulo Freire, developed what he called the "Theory of Liberation." In his book he postulated a theory of liberation for the poor that would allow them to develop socially and economically.[2] He saw the lack of education as the main reason for chronic poverty.

His teaching inspired many people all over the world to become more active and to teach writing and reading to those who were ignorant and poor. In the late 1970s, a group of six intellectuals in a large town in Mexico, inspired by Freire's theory, decided to act by building a home for street children and adolescents. Their aim was to offer "survival and protection" to the children, a home, food, and education. They soon realised how difficult it was to convince street children to come and live in the home. Street children do not trust adults, as they know from experience that adults might be keen on exploiting them in order to sell their organs or to use them for drug dealing or prostitution. In time they developed what they called *Operación Amistad* (Operation Friendship), a complex and skilful way of befriending street children with the aim of offering them to come and live in the home they had just built.

Developments

In 1998 a member of that group which had meanwhile become an NGO, *Fundación JUCONI* (*Junto Con Los Niños*, which means *together with the children*), came to London and asked the Tavistock Clinic for help. It had become clear to the workers of the NGO (who, according to the Mexican educational system, call themselves *educators* as they are not social workers or therapists) that a physical shelter, schooling, and food were not sufficient for the successful rescue of the children. Paul Freire's theory revealed itself to be insufficient. In 1999, Gianna Williams began to travel to that Mexican city in order to work as a consultant for the NGO. Williams is a former Jungian analyst who, after training in Zürich, came to London and did the child and adolescent training at the Tavistock Clinic. She worked closely with Michael Fordham introducing the method of infant observation at the Society of Analytical Psychology. She is now a very influential psychoanalyst and member of the International Psychoanalytical Association (IPA) in London. Williams has written about her work with *JUCONI*.[3]

Without using any jargon, but discussing in the group every educator's experience of each child, the interactions the educators observed including the relationships the children were establishing amongst themselves, Williams started sharing with the group a psychoanalytic way of thinking: an interest in making links and, most of all, a curiosity in attempting to give meaning to psychic pain and in recognising defences against it. Williams also introduced Bion's idea that what is perceived as an attack may be a communication; indeed, many children would run away during their stay in the children's home.[4] This teaching helped the educators better tolerate their often painful work; in fact, the turnover of staff in *JUCONI* is very limited. Gradually the educators began to observe the children with a different eye and listen with a different ear.

Williams introduced the educators to the idea of a therapeutic intervention called Special Time (ST), which was tried out for the first time by Shirley Hoxter in the seventies in a family centre in London.[5]

Special Time is a regular weekly slot of forty-five minutes of total attention offered to a child by an adult (one-to-one) in the same place and at the same time. The central skills of ST are observing, listening, and asking very few questions. Children can make whatever use they like of their forty-five minutes. Every child has a box with some material in it, pens, sheets of papers, etc., and a calendar with all the dates in which his Special Time will take place.

This method proved very useful to the educators, who were beginning to discover how important total attention, curiosity, and a non-judgmental attitude can be in the relationships they were creating with the children seen for ST. Seven educators began to offer this kind of intervention to the children living in the home.

In 2007 Gianna Williams asked me to help her supervise the educators who were offering ST and subsequently to take over for her. This was the starting point of my contribution; I began to supervise via email and Skype each session that the educators were sending me in written form. In 2008 I travelled for the first time to Mexico to meet the educators and see the children's home. Special Times were increasing in frequency, and help was needed. Spanish-speaking, qualified child psychotherapists were approached in order to meet the needs of our supervisees.

Special Time

A model was created in which each Special Time lasted one year, from February to the end of January of the following year. Children who wished to have a second, third, or fourth go had to put themselves on a waiting list to start the next round of ST with the first educator available. This system conformed to the need of the children's home to avoid too strong a transference to one particular educator over a long period of time and to encourage a transference to the children's home itself with the educators as part of the institution. Another advantage of this method was to avoid a too-close rivalry between the children over one possible favourite educator. Every child would have to adapt to another person at each round and learn to relate to each of them in a different but similar way as they all were adhering to the rules of Special Time. In a parallel process the educators were faced with a similar reality. At each round they would have to change supervisor and adapt to a new way of supervising, receiving comments written in a different style. This was possible because they too, like the children, were relating to what Williams had represented for them, a new way of thinking—all of us supervisors were representing that new way of thinking.

While at each round of ST, the children were able to represent to themselves a more coherent history of their pasts and have more realistic views about their futures, the educators were learning to bear more pain, to witness and suffer for their children in a way that just a few years previously did not seem possible. From their "protective arms" (the house) the educators began to offer the children a mental space within which emotions, feelings, and thoughts could be contained. Our team of supervisors was holding and reflecting this space in London through modern means of communications: the Internet, emails, and Skype.

This work was becoming so interesting that by 2011 I decided to liaise with the Tavistock Clinic and create a Tavistock Certificate for the educators who had never had therapy themselves and had no training in psychodynamic work. I felt that the experience with the children and the capacity to learn from the supervised work needed some form of recognition and "protection." A one-year course, "Understanding Communications with Children, Special Time," was created by me, and the Tavistock Clinic recognised it as a certificate. The manager of *JUCONI* met us in London to discuss the actual feasibility of the course, and in 2013 the course started. We were by then six supervisors.

Nine educators have now completed this course. The aim of the course has been to consolidate the educators' knowledge, offer more structured theoretical background through teaching *in situ*, and, essentially, to continue the work of supervision in written form—the educators email us the session, and we treat it as a document which has to be commented on sentence by sentence.

Running Away, Being Found

I will now discuss what have been the biggest challenges that the children and adolescents have created for Special Time and how they have been overcome.

Street children have learnt to run away. Running away becomes an internal working model, powerfully described by Symington,[6] as a way of approaching life, or a repetition compulsion, as Williams[7] puts it. From the pain of having been excluded from the love of their parents (by having been abandoned by parents who go to prison, disappear, go to work in the United States, simply leave the child with an old grandmother who has no means, or start a new life somewhere else with another partner), the children have learnt to cope by running away. Running away represents attachment to an idealised parent that "will be found" (an antidote against difficulties that is based on despair, an idealised form of manic hope). For these children, realistic hope is too dangerous, as it means giving up the fairy tale of re-finding the lost parents and accepting the reality they live in. Only this recognition will put the child and adolescent in the position of assessing what is "good enough" and give up unrealistic dreams of blissfulness. Running away becomes a perverse way of thinking according to when things don't go well, one runs away. Many of these children manically hope to reunite with their parents forever and reject what the good but limited home has to offer. Remaining attached to idealisation makes them denigrate and run away from what they have, unable to value what is really possible. I will now show how this state of mind impacts on Special Time and how it might slowly shift through the work-in-progress with the educator.

Christian (seventeen years old) ran away a few weeks after his admission to the children's home. One educator was allocated to him for his ST, but after already missing two sessions, Christian ran away with another child from the home. He stole a motorcycle, a month later was found stealing food at the market, and was taken by the police and put in prison. For certain reasons, he was put in an ordinary prison with adult prisoners. While *JUCONI* was working at getting Christian back to the home, an arrangement with the prison was made whereby Christian's ST would continue and his educator would go every week to visit him in prison, the same day of the week, at the same time.

> E: Christian, I am here to offer to continue your Special Time, but we need to agree on some points.
> C: I don't need to think about it though. I'll come anyway.
> E: How come you are so decided? In the home you never came to your Special Time.
> C: Because I understood that I did the wrong thing; I should not have left the home like that. You certainly know why I am here. ... I stole, stole a few things, then I stole a motorcycle, I bought some drugs, and so on until they caught me at the market ... and brought me here.

E: It is indeed a very difficult situation.

C: I am using here the security plan JUCONI taught me at the children's home; in case of danger do not react, ask for help. But I am not able to control myself. Look here (he shows me a big wound on his arm); yesterday they put me in isolation. Two guys were trying to take my food away from me, and I could not control myself, so I reacted and one cut me.

While at the beginning Christian seems to want to impress his educator with his "maturity," we will see how difficult it was for him to give up his belief that he can count only on himself and he cannot trust anyone.

Since that initial meeting a few months before, Christian did not attend his ST in the prison. His educator used to hear that Christian was in isolation and could not see him. Despite his good intention, Christian was, in some way, still on the run. By session twenty-five, his educator was told by the police guard that today Christian was there, but he was "crazy, impossible to reach, unworkable."

I walked towards Christian; the policeman said, "Don't let this man in, this bloke will punch him." I kept walking and came up to Christian. I asked the policeman to leave us alone. He did not move. I asked for privacy and he locked the door. Christian held his arms stretched out; he was crying, but, with what seemed to me a gesture of extraordinary courage, he walked towards me. Before I could lift my arms I understood that he wanted to embrace me. I embraced him too, and although he was still crying, I realised he was slowly calming down. He then sat down, and I asked him where he wanted me to sit. He pointed at the chair opposite him. I took his box out of my bag, and put it between us. Still crying, Christian opened his box and took out a small notebook that was inside. He began to write something on it. When he finished, he looked at me; I saw that he was not crying anymore. He made a movement towards me, so I sat down on the floor. Christian took out everything that was in his box, including his calendar with all the dates of his Special Time. He marked today's date as if it was a very special date. In this moment something powerful happened. It was as if I too became aware of Christian's Special Time, and I felt as if the supervision, the course, the reading had really changed me. I felt much more intelligent, more experienced; I understood everything, including my own feelings, my emotions, and my capacity to understand expanded.

The work of supervision had consisted of offering constant encouragement to the educator to continue to go regularly to the meetings and survive Christian's absences. The work done in supervision was a work of real hope, as we had no assurance that Christian would decide to take up his Special Time. It was not coming from a place of despair, nor of manic hope; only the tenacity of real hope—and this time we both were rewarded—brought educator and adolescent to meet in a powerful way. Some defences shifted, and the acceptance of reality brought them both to recognise each other's importance and to integrate past, present, and future. From that session onwards, working together became really possible, and, when eight months later Christian could come back to the home, a big shift had happened in him.

I will now present another vignette from the Special Time of another child, Ramón (fourteen years old), who had run away with Christian, and was found a few days later trying to steal a car. His sessions could continue in the children's prison he was put in for six months. His educator would go for every session of ST to the children's prison to give the Ramón his session. In this session we see how painful the work of acceptance can be for these children and how difficult it is to bear "reality" without idealisation or denigration. While Christian was "idealising" his capacity to survive on his own even in the cruellest prison, Ramón reacts by denigrating his family; this is a powerful defence against the pain of having been abandoned. We can see that for both children it is possible to abandon their defences only when they realise that they are held in mind (the educator does not forget them), and this is in itself a very powerful therapeutic tool in the development of a capacity to think. Experiencing another mind that can think about them has such a powerful impact on them—omnipotence and deprivation find a powerful antidote.

> R: I have many brothers, and my favourite one is called "fart." There is another one called "caca." Well, no; "caca" is my sister. Then I have another brother who is called "slag," and another one called "swine." He began to laugh, until he repeated, "yes, yes" and repeated all the names of his brothers and sister.
>
> E: I have never heard people with such names. ... It must be very painful for them to have to carry such names ... these are names used to insult people, not to call them.
>
> R: Ha ha ha, "caca," "fart," "slag" ... I have another one who is called "diarrhoea," another called "rubbish," another called "louse," another one called "pubic hair."
>
> E: I did not know you had so many brothers, and with such names ... you seem to be the only one with an ordinary name.
>
> R: It is our mother who gave us these names. ... There are more: "caw," "poop," "stupid." [And he went on and on. At the end there were more than twenty brothers with such insulting names.]
>
> E: I am amazed that at least to you your mother gave an ordinary name.
>
> R: It is all my mother's fault; her name is "*cerda*" (female pig = slut) ... yes she is called "*la cerda*" ... ha ha ha ha ha ... and do you know why she is called "*la cerda*?" Because she is a female pig! [And then imitating a pig he moved around saying,] "Yes, she is a pig and so are all my brothers and sisters."

In fact, Ramón's mother was found by the NGO and alerted about her child's state. *JUCONI*'s policy is to try to find the parents of street children and, if possible, establish a relationship between them and the child. In this case, *JUCONI* had tried to connect with her, but she did not want to know anything about Ramón, as if Ramón had never been her own child.

Later in the session, we continued:

E: It must have been very painful that she did not come at all.
R: No, it does not matter; I did not want to see her anyway. Do you think I want to mingle with a female pig (slut)? Not at all, not at all.

We can see how painful it is for Ramón to know that he is not in his mother's mind and how he transforms the pain of being the only one among her children that has been rejected by her. In his next session Ramón asked his educator if other children in the home were asking about him, if his photo was still there. Now let us see a vignette from the next session:

R: Is my photo still there?
E: Yes, it is.
R: It might be at the bottom of the panel.
E: Yes, it is.
R: So, the new children who have come to live in the house might ask, "Who is this child whose photo is at the bottom?" The educators will answer, "This is Ramón, a child who ran away and is now in the children's prison."
E: Do you think that this is what they are saying?
R: Yes. And when I come back I will tell them, "Look, I am back again; don't run away, it is not good."

In order to survive, Ramón needs to destroy his own mother (and his brothers and sister) with whom he cannot imagine being happy and close, and instead imagine that he will manage alone to survive the prison and come back to the children's home like a hero, as is depicted in the following:

R: I know that when I come back to the home everybody will be waiting for me. Those who already know me will say [his face lifted], "Welcome back, Ramón!" All the educators will be there, telling me, "Welcome back, Ramón, darling, this is your home!" And I will begin to smile at everybody, and I will say, "Thank you, thank you." And I will kiss everybody [he was making faces as if he were rehearsing the event], and the new children will say, "Ah, this is Ramón, but he does not look like the photo; how

come?" And the educators will answer, "Well, he has spent a few months in the children's prison; he has grown, matured. He has suffered a lot."
E: It seems to me that you have been thinking a lot about your re-admission.
R: Yes, everybody will be happy to see me again.

On the other hand, we can see how the concept of a mind with space for him is really occupying Ramón's mind, and how, while reflecting about his experiences, he is beginning to develop a mind of his own.

A few sessions later (five months into his ST), Ramón begins to discuss his death with his educator, and he seems very preoccupied about being remembered then. This persistent thought will be the red thread of the second part of his ST. Ramón is really trying to make sense of what it means not to be in anybody's mind, to be abolished; only death comes close in his mind to what his experience has been. It will take many more rounds of ST for Ramón to be able to have a better sense of this in relation to his mother, himself, and others. Nevertheless, three sessions before the end of his first round of ST, when he is back in the home and has to separate from his educator, he begins to talk about the terrible pain of not living with his family.

Double Deprivation and Omnipotence as a Defence against Survival Anxiety

Children who are deprived of parents who can look after them physically and mentally tend to develop strong defences against the pain of having been deprived of what is archetypally expected. Survival anxieties give rise to omnipotence, the child becomes the invincible hero; and to double deprivation, the child tends to reject what he has and goes in search of something that in his mind is better and promises total happiness. These children cling on to an internal object that promises them blissfulness. Equipped with these defences they become impermeable to fear, invincible, and without needs. When these defences begin to crack, the mental pain the educators and children have to bear is very heavy, and the work of supervision becomes, for the educator, an important refuge and nourishment.

Esteban

I will now present a brief résumé of four years of ST with a child called Esteban. My aim is to show how, from omnipotence and double deprivation, Esteban was able to begin to talk about mental pain, fear, and anxiety.

Esteban was eight when his old grandmother died, and for some reason he was brought to the children's home by a neighbour who had heard of *JUCONI*. Esteban has never been a proper street child, but his life was a life of neglect, abuse, and extreme poverty. His mother left his father when Esteban was one. She left with her three older children because her husband was an alcoholic and hit her regularly. She decided not to take Esteban with her because he was too small, and perhaps she left him with his father out of guilt; at least he could have one child to look after. Shortly afterwards, his father brought home woman after woman, until Esteban was five, and the next woman stayed for two years. During these two years he was sexually molested by her, and his father went to prison for stealing from a shop. Esteban went to live with his grandmother who was extremely poor and eventually died one year later. He went to school from the age of six but by the time he was moved to the children's home he could barely read and write.

Four Years of Special Time in a Few Words

In the first round of Special Time, Esteban used to masturbate—possibly a sort of "sexualised thumb-sucking." In this way he did not need to relate and he was locked in his own world, soothing himself. Only slowly did Esteban begin to play with toys, and, by the end of his second year of ST, he enjoyed playing with a penguin who "lived on his own in the north pole." In this way Esteban was showing how he used to survive emotions and make himself impermeable to suffering and pain. Close to the end of the second year of his stay in the children's home, his father agreed to see him and Esteban began to meet his father in prison regularly on a monthly basis, accompanied by an educator. This reality, together with the life in the home broke down the penguin attitude of Esteban, and a chaotic picture of his internal world emerged in the sessions of Special Time; *La Parca* (a dressed skeleton worshipped in Mexico as a deity) began to fill his sessions together with policemen, doctors, the spider man, children, criminals, and teddy bears in constant and endless fights of destruction.

In his third round of ST, Esteban was allocated a male educator (in years one and two he had two different female educators), and he began to spend most of the time playing football in the sessions. His aggressiveness, rage, and desire to compete and win were contained in fighting with his educator who remained stronger than him and was able to contain Esteban's emotions in a physical way. It was towards the end of the third year of ST that Esteban began to show some fear of his own deepest emotions; he acquired the habit of going to the toilet during meaningful moments in the session, spending some time there. It was when his educator began to address these "mini-running-away attempts" that Esteban could begin to talk about his life, his father, his memories of the past, and his great fears of the future. At the end of his third round of Special Time, Esteban was eleven years old. He took a break from ST and resumed with the fourth year when he was thirteen. His educator, in her detailed session reports, observed and interacted with a child who had changed a lot: a young adolescent full of fears and anxieties about his future, including his fear of becoming alcoholic like his father; his difficulty in imagining a future outside the children's home; and his desire to reunite with his father once he came out of prison. By the end of that year, Esteban had developed a more coherent sense of himself in the world. Naturally, the work offered to the children in the children's home helps enormously with their psycho-physical development; all the more, the attention encountered during Special Time offers the child a space in which their internal world emerges, is given particular attention, and subsequently finds expression and meaning.

It is a powerful experience to see that although these children have not known what we would call "a normal up-bringing," they can develop what is normal and to be expected in a human being: a capacity to feel and suffer, to think about themselves and others in a coherent way.

Trauma and Repetition of Trauma

It is not always possible to reconstruct the past of these children. Sometimes there is no one who remembers it. We know that humans tend to repeat their traumas in unconscious enactments, and that these repetitions have two sides; on the one hand, the trauma serves as a basic experience around which our deepest experiences of identity are formed, and on the other, repeating a trauma might have some value in trying to understand and order it. It is beyond the scope of this chapter to go into the details of this problem; nevertheless, in the following story I would like to show another way of understanding "running away"—as a repetition of a trauma that has happened but cannot be remembered.[8]

Alberto one day disappeared. He had lived in the children's home for a few years, had had three rounds of Special Time, and was well adjusted in his development. Alberto was now seventeen, had learnt a practical skill, and was in the process of looking for an apprenticeship. He could live in the children's home for another two years. The educators of the home were puzzled by his disappearing one day, and it was only after many months that Alberto was found. The story he told was the following:

> One day he found a job as a builder for a man who took him in his car and showed him what he had to do. This man left Alberto in a region of the province far away from the city. Alberto did not feel able to run away because he found himself "lost," incapable of orientating himself. He stayed there and worked for the man who did not pay him and brought him little to eat.

It was only when he was found by *JUCONI* in extraordinary circumstances, and after he had recovered from what had been a sort of kidnapping, that Alberto could emerge from a sort of dissociation, suddenly remembering what it was like when his parents had left him with a neighbour before they disappeared. He was five. They never came back. Eventually, he ran away and later went to live in the children's home. He had lost his memory of that event. In his subsequent Special Time (his fourth round), Alberto was able to reconnect with that lost memory, and the past experience could be reintegrated in his life history and worked through.

Separation

Separation is the most difficult aspect of ST. Children and educators are afraid of suffering separation; it has been a very complex task in supervision to guide educators in preparing the children well for the end of their Special Times. The fact that there are so many rounds of ST for each child facilitates the experience of separation, and its re-working brings about old conflicts and memories, as well as new insights in the children, educators, and supervisors.

In my first visit to Mexico I did some observations of the children in the home. In particular, one child drew my attention. I did my observation in the classroom the children used for their homework. Normally, there is an educator to supervise them and help them with their questions. I observed that every time the educator had to leave the room for some reason (get a pen, call another educator, etc.), this child would lie down on the floor in the foetal position, and as soon as the educator came back, the child would stand up again, able to continue with his task. In one hour the educator had to leave the room at least six times, and inevitably the child would lie on the floor and then stand up. In my mind, trying to make sense of my observations, I thought that the educator represented the spine of the child, and without his presence, the child would collapse like a baby, unable to hold himself up. I found it very interesting to see the child jump up again as soon as the educator came back, as if his presence helped the child to find the external support he had not yet developed internally. The comings and goings of the educator were working as mini-separations and mini-re-encounters, which, for the child, were so helpful. For most of the children, leaving means "abandonment." The separation from each round of ST at the end of each year helps the child to integrate endings and new beginnings and serves the purpose of learning what ordinary children know—that a good separation is when the child has internalised the presence of his parents and can survive without them. The fact that Special Time is taking place in the children's home allows the child to separate from it while still being held by the home that offers containment in the transition from one round of ST to the next.

THE INSTITUTION AS A FAMILY

In this chapter I have tried to show how much work has been necessary for a home for street children to act as the "parental couple" in order to offer what for other children is ordinary. The NGO, the fundraisers, the educators, and our team of supervisors, including Gianna Williams' input, have acted as the "parental couple." We have offered our cultural tradition (psychodynamic) as a link between the maternal and paternal functions of the home. This link has created a space for growth. As happens in families, children leave and go to create their own families; by now the oldest children of the children's home have left, but they approach *JUCONI* asking for help in the upbringing of their own children. *JUCONI* is setting up different kinds of interventions with preventive work in mind. At the same time, those educators who have completed our course are now beginning to teach educators of other institutions how to set up Special Time, and I, using the Mexican experience, am setting up a similar certificate for psychologists working with children in Russia.

I would like to thank Sandra Cortés-Iniesta and José Francisco Meregalli for allowing me to use some vignettes from their Special Time case material.

NOTES

1. Donald Woods Winnicott, "The Theory of Parent-Infant Relationship," in *The Maturational Process and the Facilitating Environment* (Madison, CT: International University Press, 1965), pp. 35–55.

2. Paulo Freire, *The Pedagogy of the Oppressed*, trans. Myra Bergman Ramos (New York: Continuum, 2000).

3. Gianna Williams, "Work Discussion Seminars with the Staff of a Children's Home for Street Children in Puebla, Mexico," in *Work Discussion: Learning from Reflective Practice in Work with Children and Families*, eds. Margaret Rustin and Jonathan Bradley (Karnac: London, 2008), pp. 253–66.

4. Wilfred Bion, "Attacks on Linking," *International Journal of Psychoanalysis* 40 (5–6, 1959): 308–15.

5. Shirley Hoxter, *The Old Lady Who Lived in a Shoe* (London: Tavistock, 1981).

6. Neville Symington, "The Survival Function of Primitive Omnipotence," *International Journal of Psychoanalysis* 66 (4, 1985): 481–87.

7. Gianna Williams, "Thinking and Learning in Deprived Children," in *Internal Landscapes and Foreign Bodies: Eating Disorders and Other Pathologies* (London, Duckworth & Co., 1997), pp. 25–31.

8. Alessandra Cavalli, "Transgenerational Transmission of Indigestible Facts: From Trauma, Deadly Ghosts, and Mental Voids to Meaning-Making Interpretations," *Journal of Analytical Psychology* 57 (5, 2012): 597–612.

Chapter 15

Returning to China

John Beebe

First Return

Arriving in China in December 1998 to deliver the keynote address for the First International Conference on Jungian Psychology and Chinese Culture at South China Normal University in Guangzhou, I was encountering China again for the first time in fifty years. As a boy of seven and eight, I had lived in Nanjing where my father, a Major in the U.S. Army, had served the last American mission to the Kuomintang government as an assistant military attaché. The years I spent there reached their culmination in June of 1948, when my mother had root beer flown in from the States by military transport to serve at my ninth birthday party.

It was during that year that a Chinese girl of about twelve, whom I passed in walking home to my family's large house, looked me over and remarked in very careful English, "You are a pig." Nothing ever said to me in my life has made a deeper imprint. At that moment I felt my status as a privileged citizen of a nation that had, in recent Chinese memory, aspired to be a colonial power, or at least, as my father would say, to "influence" the future of China. When my mother said, "Communism will be good for China," I chose to believe her.

She and I were "evacuated" that November, in the face of an imminent invasion from the North by Mao Tse-tung's insurgent forces. By that time, she was bent on divorcing my father, who remained in China even after Nanking, which had been Chiang Kai-shek's capital, fell to the Communists. My father would be under house arrest for an entire year before he was allowed to return to Washington. By then my mother had divorced him and I was already hardened in my opposition to his anti-Communist stance.

Reacquainting myself with China in 1998, I saw a country more in the grip of Westernization than it ever seemed in the days of the Kuomintang. Banners for Colonel Sanders' Kentucky Fried Chicken hung in the dark air on what seemed like every other street corner. More than twenty years had elapsed since the death of Mao. The conference's Chinese hosts were ready to welcome to China not only a few psychological visitors from the West but, with us, some of the ancient culture and ideas that the country had dedicated itself so fiercely to uprooting in the late 1960s and early 1970s.

At that conference, I could sense that the group of people brought together by Shen Heyong, principal convener of the conference and a professor offering comprehensive courses in personality theory at the university, was excited to engage with Jungian psychology. Nevertheless, it was not at all obvious what form their engagement would take. At that time, there was almost no psychotherapy in China other than counseling services offered to university students. A few newly trained psychiatrists were at the conference, but they told me that their practices were almost exclusively pharmacological.

The culture was still very collective. Many of the people I met, especially the young people, were more comfortable thinking of themselves in their various group-member identities than in considering themselves individuals with unique gifts and aspirations. During the conference, a Chinese psychologist who had studied in the U.S. and was now working in student health in a university explained that he had administered the Minnesota Multiphasic Personality Inventory to a representative sample of Chinese students and determined that a third of them were clinically depressed. It was not hard for me to imagine why. The students lucky enough to qualify for university positions as a result of competitive examinations could look forward to years of very hard study in cramped, under-heated quarters in order to qualify for good jobs in Western franchises springing up in a country for rent to the West. The all-important ingredient of a young person's education—something to idealize—was missing in the new materialism informed by the "One Country, Two Systems" revolution spearheaded by Deng, which replaced a perhaps too literal belief in the possibilities of Marxism with a belief in nothing at all but material success.

This was not so, however, among Dr. Shen's students in Personality Psychology at South China Normal University. They really believed in the power of personality psychology to bring change. They wanted to hear how analytical psychology could return some of China's spiritual values, encoded in the *I Ching* and the three great "ways of thought" that had emerged to clarify that book's understanding of what it means to stay in *tao*: Confucianism, Taoism, and Buddhism, each of which had been studied by C. G. Jung and, in significant, still recognizable ways, incorporated into his psychology of how a personality develops.

I didn't think of it at the time, but I brought to that first international conference on Jungian Psychology and Chinese Culture some of the same libido that I might bring to a first meeting with a new analysand, in that, I made an effort to demonstrate a capacity for creative listening and for a kind of mirroring that reflects not only the individual's present reality but also the future potential that might actually show itself more clearly to me than to them.

In this spirit, even though a number of the Chinese audience members expressed mystification and disbelief that there could possibly be anything substantive to talk about in individual analysis on a weekly basis, I set out to convey a sense of what Jungian analysis might be like, not at the level of counseling, where they ostensibly might have more readily accepted it, but at its most powerful, symbolic, and transcendent.

Here is some of what I said in that talk.[1] I began by calling the attendees' attention to J. J. Clarke's book, *Jung and Eastern Thought*[2] for its tracing of Jung's connection to Chinese, Japanese, and Indian thought and culture.

> Much of Jung's emphasis was on Chinese culture; his interest in Buddhism was evident from the very beginning of his student days, when he was reading Schopenhauer,[3] to the last year of his life, when he was reading the Buddha's discourses again.
>
> In the middle of this lifelong interest in Chinese thought, Jung talked about the *tao* in the book *Psychological Types* and about *wu wei* in his Commentary to *The Secret of the Golden Flower*.[4] Above all, he made sure that the Richard Wilhelm translation of the *I Ching*[5] was also translated into English and produced in the American book market—I would say

personally that Jung's Foreword[6] to the *I Ching* is my favorite of all his writings.

Today, I do not want to dwell on this intellectual history—of which I think Jungian psychology can be very proud—but rather to say how we approach everyday problems of doing psychotherapy, how it actually lives in our work and lives, and therefore how Jung lives in our work and lives.

Just as it is very difficult to get a sense of what the great philosophical figures were like, what Lao Tzu was like, what Confucius was like, and above all what Buddha was like (and we get a better sense of what these people were like by getting close to some of the people who followed them and have incarnated their spirits), so I would like to tell some stories that come from my experience with my own teachers, since both of the teachers that I'm going to describe had personal contact with Jung and Jung's way of seeing things. I believe that particularly, Jung's connection to Chinese thought influenced the way these teachers approached the problem of being doctors, healers, and psychotherapists.

Now, to become a Jungian analyst, one has to undergo the experience of being in psychotherapy with a Jungian analyst. One of my first Jungian analysts was a woman named Elizabeth Osterman. When I came to her, I was a young psychiatrist, just finishing my residency training at Stanford University. At that time, I was very impressed with medicine, with being a doctor, and with what I had learned to be a psychiatrist, and I was interested in the power of psychiatric medicine to change the human mind.

Therefore, I was interested in knowing how my analyst, Dr. Osterman, had approached the problem of becoming a psychiatrist and a Jungian analyst. And she related an interesting story to me. Originally, she had trained to study viruses; she was a microbiologist. And so she was already a professional woman when she first felt a need to go into Jungian psychotherapy for herself. In fact, she had had a dream in which there was a woman lying on a bed who was very old and sick and tired; when she had this dream, she knew immediately that the woman was in herself and was in some degree of trouble and that she would need to go to a psychotherapist to figure out what the trouble was.

She had a successful analysis, and as part of the analysis she completed studies to become a medical doctor, and went forward with the idea of becoming a psychiatrist. So, in our city she went to a great psychiatric institute, which is notable in its architecture because it's built a little bit like a tower; it has a rather tall structure. You could call this a pagoda of psychiatric learning. Remember this is already a formed adult woman who has had several years of Jungian analysis and she's just about to start to learn to be a psychiatrist. The night before she goes to the institute the first time she has the following dream:

She is looking at a building that has a tall tower, and at the top of the tower, right at the rooftop, a man and a woman are wrestling. And they wrestle, and they wrestle, and they wrestle, and they wrestle, all day and all night, and they just go on wrestling and wrestling.

Now it happened that when Elizabeth had this dream, she had the good fortune that that day passing through San Francisco was one of Jung's most famous colleagues, Dr. C. A. Meier, who wrote many books. And so she told Dr. Meier her dream, and she said, "I was looking up at the psychiatric institute building, and there was this man and woman wrestling on top of the roof all day and all night long." And Dr. Meier said, "Ja, and they never come down."

So Elizabeth had to come down on her own. In those days in California where, as in every other place in the world, sooner or later one has to have a license to practice any form of healing, one could get away with going to a psychiatric residency for as short as one year. On the strength of that dream she went for only one year and then got away, because she felt that it would threaten her as a woman and as a psychological person if in that place of great psychiatric learning this problem of the male and the female opposites was going to be wrestled with so much up in the head, so high up, and so intellectually, and never, ever come down to the ground of being.

Elizabeth never tired of telling me and telling all of us in our Institute who knew her that what Jung was talking about was Nature. I believe there's a very wonderful term for this in Chinese culture. It is Earth. You know, you have the three terms *Heaven*, *Earth*, and *Man*. In thinking of this dream of the male and female wrestling with each other so high off the ground, we have to think of their distance from the Earth.

> Now, we know Jung has talked a great deal about the *tao*, and I think that one has to understand the *tao* as the watercourse way that passes through the earth. In Hexagram 2 of the *I Ching*, which is the Earth hexagram, we have the wonderful line describing the nature of *tao* itself: "Straight, square, great. Without purpose, yet nothing remains unfurthered."[7] This seems far away from the male and female opposites struggling on top of the tower. Jung did not like Western philosophy when it got too much into an endless argument. He felt it was better when it was Nature, or what he called natural philosophy. And this is what Elizabeth had to teach me as a young psychiatrist again and again when I would struggle with my dreams and try to understand what they mean—and tell my feelings what they were doing. She saw my dreams as expressions of Nature struggling to get through to me.

I then turned to a story about Joseph Henderson, who had also been my analyst.

> When I came to work with Dr. Henderson—who founded the Jung Institute of San Francisco and was one of Jung's most long-standing and loyal followers, having known Jung as far back as 1929—I'm afraid that I projected onto him, as we say in Jungian psychology, the very figure that appeared to Richard Wilhelm in the dream that he had shortly before he undertook to translate the *I Ching*. At that time, Richard Wilhelm was in Tsingtao, and he dreamt that a beautiful old man appeared to him who said that his name was Mountain Lao, and that he would teach Reverend Wilhelm the secrets of the mountain.[8] So when I first met Dr. Henderson I'm afraid I saw him too as Mountain Lao, and I said to him, "It seems to me as if you've always been here."
>
> Dr. Henderson is not the usual American man. He has a very English *persona* (the word we use in Jungian psychology)—that is, he has a very formal way of presenting himself that almost seems like it belongs to a past time.
>
> So, when I first came to Dr. Henderson, in that first hour with him, I looked at him and said, "What I like about you is that you have never cared about what other people thought." Dr. Henderson said, "Well, I have always had an independent mind." And then he cleared his throat and tried to say something else, but his tongue got somehow tied up in his mouth, and he could not say what he wanted to say. He cleared his throat and

tried to speak again. And his tongue got curled up in his mouth and he could not speak again. I went right on talking, as if nothing had happened. And when I had finished my next too many words and there was just a breathing space, Dr. Henderson looked at me and said,

The reason I become tongue-tied just now is that it really isn't true. I actually cared very much what other people thought all my life, but I had to go away to school, to a boarding school far from my home when I was young, and to protect myself I had to develop a face that didn't show emotion [what we call in English]—a poker face. And I got rather too good at it.

Now imagine that this is the beginning of an analysis! Notice that Dr. Henderson did not hide behind the projection of Mountain Lao—eternal, the carrier of the wisdom of Heaven, beyond being touched—that I had placed upon him. Notice also that he *tried to*, at the beginning, when he said, "Well, I have always had an independent mind." But notice that his body would not let him get away with that. And notice that he noticed that himself.

The part of the mind, for a Jungian analyst, that will not allow us to proceed if we're living in a fiction is precisely what Jung referred to as the *tao*. The choice we make to honor that, when it appears in us, is what I think we should call the *te*. And what was so impressive to me in that first hour with Dr. Henderson was that he lived that relationship for all that it was worth and did not care about the appearance of authority when he did so. This meant that if I was going to have what in analysis is called a transference, that is a projection of the possibility of development of the Self (an archetypal transference is the Jungian technical term), if I was going to have this, I was going to have to have it on the basis of who Dr. Henderson really was, not on the basis of a fantasy of some unreal person that he could never be.

You see, I had come, as I think many people come to psychotherapy, with the fantasy that my depression and my anxiety—what we call in the West my vulnerability—was going to go away as the result of analysis, the way that depression goes away if one takes an antidepressant, or the way a headache goes away if one takes an aspirin. When Dr. Henderson insisted that however he looked from the outside, he, too, was just as vulnerable as I was, I suddenly realized that my vulnerability *was* me, and it would never go away. I believe that Dr.

Henderson learned this from Dr. Jung. And I believed that Dr. Jung learned this from Lao Tzu.

The Chinese audience's reaction to the talk I have excerpted here was interesting. Many of the younger audience members who had grown up in the shadow of the Cultural Revolution were shocked to learn that I had spent so many of my years making sense of key elements of Chinese culture that the first half-century of the People's Republic had tried to modernize away.

Since the publication of Jung's *Red Book* was still a little more than ten years away, we had not yet encountered the power of Jung's writing about what we owe the dead,[9] but in retrospect each of the Jungians who participated in the 1998 conference was not only bringing new ideas and techniques to China but also encouraging the Chinese audience toward a new, appropriate relationship to their ancestors and legacy.

Turning the Light Around

In the years that followed, I returned to China for several more international conferences, witnessing enormous economic and cultural changes. One of the projects that had been raised at the 1998 conference—to start a dialogue between the western Jungians and the Chinese about what Jung had borrowed from Chinese culture—gained strength and mutual interest. On the one hand, we analytical psychologists were helping to awaken an interest among some of our Chinese interlocutors in the *I Ching,* Chinese alchemy, and certain principles of ancient Chinese philosophy—subjects that had been off the table in China for many years. On the other hand, we were also forced to begin to consider that some of analytical psychology's borrowings from China may have been based on incomplete understandings or, in some cases, even on creative misunderstandings. We were returning to the Chinese some of their cultural inheritance, but they were giving us new ways to think about what we had considered settled concepts.

At the Third International Conference on Analytical Psychology and Chinese Culture, in 2006, in a talk entitled "Individuation in the Light of Chinese Philosophy,"[10] I focused on Jung's conceptualization of the individuation process and the relationship of that concept to what Jung knew of Chinese philosophy:

A process with this name (individuation) would imply, perhaps, the flow of insight in a rich experience of the soul that psychological consciousness can promote, but it does not imply a shift in the center of that consciousness, as does "turning the light around"—a process that is the basis or beginning of the Golden Flower practice and which eventually leads to the discovery of the heart of the mind. What is reached by this process is *fu*, a place in the mind where thought itself arises and where the mind can be contemplated objectively and can also operate objectively. The mind in this place is spoken of as a mirror and experienced as ego-less. Attaining *fu* involves a shift from the ego-centered, subjective use of mind toward an objectivity with regard to the thinking and feeling processes themselves. Such objectivity is the most salient characteristic of a mind that is grounded in what Jung calls the Self. It is this shift of emphasis that Cleary attempts to capture in his translation of the Chinese phrase *fan-chao* as "turning the light around."[11] The idea is beautifully conveyed by the Tang Dynasty poet Wang Wei's well-known four-line poem about being in a deer forest:[12]

> There seems to be no one on the empty mountain ...
> And yet I think I hear a voice,
> Where sunlight, entering a grove,
> Shines back to me from the green moss.

The "light" in the natural metaphor is consciousness, and "turning it around" is meant to convey not a circuit but a reversal of the subject-object polarity on which ego-consciousness depends. When the light "shines back to me," the subject becomes the object. My former egoistic use of consciousness is now observed. And observed, my consciousness is deflected and re-emerges as a light grounded in nature, in the experience of what Jungian psychology calls the Self, Buddhism the "Buddha-nature," and Taoism "the source before father and mother." This is a less-directed light, but one with a lot more affection for the real and more potential to get me to see than before.

Without practical experience, one cannot understand what is being talked about, but I would point out to those who have never taken up Chan Buddhist or Perfect Realization Taoist practice, or wandered through the forest of the Self in a long Jungian analysis, or reached the "evening light" of spontaneous individuation after age forty-seven, that something like this reversal of polarity occurs every time we go to sleep and dream. Then the perspective of the waking ego, with its lusts, attachments, and anxieties, is met in a variety of symbolic guises and becomes the object of another perspective that somehow remains outside the subjectivity of the anxious ego. The ego continues to be observed and felt in the dreaming state, by a self that is at least somewhat detached, but we see the ego as a shadow in our dreams. Our dreams invite us to critique the ego's standpoint. They present things that the ego is unwilling to face in the waking state. Therein lies the tremendous power of dreams, since during the day, the ego's standpoint is more or less thoroughly identified with, so much so that even when the ego is anxious, it is somewhere almost always pretty sure that it is right and believes its troubles are a consequence of the lack of sympathy of others. In a dream, on the other hand, even at the height of anxiety, the self that is dreaming may seem to identify with the perspective of the frightened ego, but there is self-awareness that wonders if maybe the ego's state of mind is inauthentic, hyperdramatic, or absurd. In a dream, the thought, "How did *I* get mixed up in this?" may come.

In other words, the dream makes clear that the Self is unhappy with the ego's perspective on things. It's that ability to turn the light of any consciousness around and examine where it is coming from that makes the dream state so special and so remarkably effective in promoting reflection on the ego's perspective. In the waking state the perspective of the ego is less often questioned unless the person has been trained in questioning the ego, as through Buddhist practice or through a psychoanalysis that invites the ego to look at its shadow.

Jung seems to have grasped, through his amplification of images in the text, that the Chinese meditation manual was talking about the move from ego to Self. One of these images was the Golden Flower, which Jung recognized as a mandala image. The text induced Jung to amplify the flower in a Buddhist way. The flower is an image that the Buddha himself resorted to more than once to explain enlightenment to his disciples. In his famous "Flower Sermon," when, near the end of his life, he found himself still trying to express to his followers, gathered by a quiet pond, the essence of his teaching, the Buddha pulled out of the water a golden lotus flower, with its roots intact, and held it up for them to see.

There is an analogous, though more dialogic moment between Jung and a student in his June 22, 1932 English seminar on Visions. The questioner, the Swiss Fritz Allemann, had asked, "Is not individuation, in our sense of the word here, rather living life consciously? A plant individuates but it lives unconsciously." Dr. Jung answered,

> That is *our* form of individuation. A plant that is meant to produce a flower is not individuated if it does not produce a flower, it must fulfill the cycle; and the man that does not develop consciousness is not individuated, because consciousness is his flower, it is his life, it belongs to our process of individuation that we shall become conscious.[13]

I followed with examples from the *I Ching*, from the father of neo-Confucianism Chu Hsi,[14] and from the Taoist humorist sage Chuang Tzu,[15] each illuminating an aspect of the way the Self sheds its own light on matters of concern to the ego. I was letting the Chinese see what I had learned from their theoretical traditions about my own, and thus how China had retained the power to affect an analytical psychologist's attitude toward individuation. In a reciprocal way that Confucius would have loved, many Chinese people, after this lecture, wanted to take my picture.

The *Ding*

It was not until three years later, after I had spoken at the fourth international conference, that Chinese participants wanted to be included in such pictures alongside me. This time, my theme was entirely Chinese. My lecture was entitled "Achieving the Image."[16] Throughout it, I projected a photograph of a Shang dynasty *ding*, a large bronze caldron with two ear-like carrying rings that I had stood in awe before at the Shanghai Museum, just a day or so before giving the lecture. I introduced the image in the following way:

> In the course of writing a Foreword to the English version of Richard Wilhelm's translation of the *I Ching*, Jung decided to ask the Book of Changes how it felt about his intention to present it to the Western mind. As is well known, Jung received Hexagram 50, "The Caldron," which in Chinese is *ting* (or in pinyin *ding*). Roderick Main has recently commented that "the nature of the *I Ching* ... may indeed be appositely symbolized by the *ting*, the cauldron or vessel," since Wang Pi had already written in the third century CE, "The Caldron is a hexagram concerned with the full realization of the potential in change."[17] Main adds that a later questioner mentioned by Stephen Karcher and Rudolf

Ritsema in the introduction to their own translation of the book asked the *I Ching*, "Who are you, how shall I use you?" and got Hexagram 50 as well.[18] These translators translate the *ding* as "The Vessel/Holding," and they relate its many meanings as container and ritual cook pot to the "imaginative process" that enables human beings to metabolize their experiences through the use of symbolic formulations of it.

The image of the *ding* has managed to stay vital for eighteen hundred years as a description of the *I Ching's* uncanny way of "cooking" our life experiences for us in such a way that we can digest them. Professor Liu has just told us that Dr. Shen received this hexagram as the starting point of his answer when he put the question to the oracle of how analytical psychology might develop in China.[19]

One can see Shang dynasty *dings* in any good collection of Chinese art; so we know the image is deeply rooted in Chinese culture. This ritual cooking pot, as Ritsema tells us, usually has "three feet and two ears." We can pause for a minute to think why such an image would have become the standard depiction of a "sacred vessel used to cook food for sacrifice to gods and ancestors." Are the three feet the great Triad of Chinese thought: Heaven, Earth, and Man? Are the two "ears" of the vessel the yin and the yang—the masculine and feminine ways of receiving experience when life tries to send its messages to us? No doubt, but is defining these separate features of the *ding* in such an analytic way really more powerful than simply contemplating the unified image of the Caldron? Was it really conceived with these allegorical features in mind? Or is its visual combination of hollowness, a triadic base, and two ears by which to lift it to ourselves or hold it out to others not the way the image gets across the practical uses of holding, ritual, and vessel?

My conclusion was that the same image, the *ding*, presented itself to Wang Pi, Jung, and to Ritsema's anonymous questioner as a metaphor for the *I Ching* because the image was the one best suited to the purpose.

> That's one definition of archetype—an image that is not just ancient, but accessible to all. Getting such an image to present itself requires relatively little work, because the image is already out there, waiting to be accessed. Other archetypal images can appear only through a more creative process of emergence, arising from the ability of the unconscious to

synthesize images freshly. This is a more subtle understanding of archetype that has been advanced by Joseph Cambray[19] and others. To access either kind of image requires what the *I Ching* calls modesty; to let it in, one has to suspend one's own image-finding capacity and instead allow the unknown thing one is trying to grasp to supply its own image of itself, whether found or freshly created. Let us say, then, that the first step in achieving any image with the archetypal power to transform our understanding is to suspend conscious attempts to symbolize the thing we wish to understand. Only then can the image of the thing, whether already there in the unconscious or in need of its own emergent process to attain form, come forward. We have to practice *allowing the image to present itself*. That is the wisdom in the standard Jungian practices that so many people find dubious—accepting images presented to us by dreams and divination.

I think this talk, which made it clear how much I had accepted Chinese images, helped me to begin to engage more deeply, in an equal, mutually embracing way, with people in China. This shows in the many photographs taken at this conference in 2009, in which I am holding and being held by Chinese people with whom I was now sharing analytical psychology in a Chinese way. It also shows what my life has turned out to be in the past five years, after being made Liaison to the International Association for Analytical Psychology (IAAP) Developing Group (DG) in Shanghai, and then Visiting Analyst to the Guangzhou Developing Group, and finally Visiting Professor of Analytical Psychology at the City University of Macau; I have become analyst, supervisor, and teacher to literally scores of Chinese who have found it easy to talk to me about their lives and who have been able to take aboard some of the most elusive concepts in analytical psychology.

I have been fortunate enough to teach for each of the IAAP's DGs in China, including Hong Kong, Taiwan, and a not yet official group in Beijing, and, by Internet to people tuning in from Guangzhou, Shanghai, Beijing, and Macau, a weekly two-hour supervision seminar by Skype. Originally the *Shanghai Seminar*, I think of it now as the *Synchronistic Seminar*, as it is my chance to model thinking psychologically in the moment. The topic of each seminar session is set by a student's question, offered to me without my knowing what it will be.

These days I am visiting China once or twice a year to teach in universities and to psychotherapists at various stages of training. Each week I give seven hours to see patients by Skype (and in person, when I am in China). We both consult dictionaries, and I sometimes type both the dreams people tell me and some of the amplifications in the instant messaging box Skype provides. It is remarkable how extraordinary the face-to-face exchange can be that the close-up medium of Skype provides in an era when people have finally begun to value the television image. Several of my analysands have been Router Candidates in the IAAP, and I am struck by how amazingly alive to psychological matters they tend to be. Our exchanges are suffused with the joy in the work that can accompany dealing with even the most painful topics. Every aspect of Chinese life emerges, but there is little tourism in the way I am doing therapy. Rather, we are constantly proving that the collective unconscious is common to all and that the complexes of another culture are not unlike those of our own in being complexes. As a consequence, I am learning as much from my Chinese patients' work with me as they are from mine with them.

I wouldn't have expected a practice that now keeps me up late at night seeing patients in a time zone fifteen or sixteen hours in advance of my own, conducted between people who have such different native languages as Chinese and English, and by electronic means linking me to people six thousand miles away, to be so invigorating. Yet it has changed my life and the attitude I hold toward Jungian psychology. A line from the *I Ching*, in the Hexagram Inner Truth, says,

> A crane is calling in the shade.
> Its young answers it.
> I have a good goblet.
> I will share it with you.[20]

The crane, in China, is a symbol of longevity, and it does seem to describe me at seventy-five, craning to hear what someone, usually thirty or forty years younger than I am, is telling me in somewhat broken English pronounced in a distinctly Chinese way. I am sharing what I have in my goblet, but it is also rejuvenating me. I am also the young person in the very old Chinese culture, just as I was as a boy of eight, except the boy, now, is conscious of the gift he has received, and despite the delicious food he is actually served, perhaps no longer quite such a pig. He shares with his students.

I foresee that many Jungian analysts around the world will allow themselves in the coming years the pleasure of returning to China some of what analytical psychology originally learned from it. If their experiences are at all like mine, they will be enriched and enlivened by engaging with the Chinese Jungians who have already done so much to refresh their own cultural traditions. One particularly gratifying aspect of this work has been to help the new Chinese therapists become freer to use their own self-experience in their work as counselors and healers.

Notes

1. John Beebe, "Keynote Address from the First International Conference on Jungian Psychology and Chinese Culture, South China Normal University, Guangzhou China (1998)," *Quadrant* 31 (2, 2001): 19–30, esp. 20–24.

2. J. J. Clarke, *Jung and Eastern Thought: A Dialogue with the Orient* (London: Routledge, 1994).

3. "One of the Most Important Influences in the Shaping of Jung's Interest in Eastern Philosophy," in Clarke, *Jung and Eastern Thought*, p. 59.

4. C. G. Jung, "Commentary on *The Secret of the Golden Flower*," in *The Collected Works of C. G. Jung*, vol. 13, ed. and trans. Gerhard Adler and R. F. C. Hull (Princeton, NJ: Princeton University Press, 1929/1967), pp. 1–56; see § 20.

5. Richard Wilhelm, *The I Ching or Book of Changes*, trans. Cary Baynes (Princeton, NJ: Princeton University Press, 1977).

6. C. G. Jung, "Foreword," in Wilhem, *The I Ching or Book of Changes*," pp. xxi–xxxix.

7. Wilhelm, *The I Ching*, p. 13.

8. Richard Wilhelm, *The Soul of China* (Maple Shade, NJ: Lethe Press, 2007), p. 180.

9. C. G. Jung, *The Red Book: Liber Novus* (New York: W. W. Norton & Co., 2009), p. 297, note 187.

10. John Beebe, "Individuation in the Light of Chinese Philosophy"; I am quoting here from the slightly edited version of this talk that appeared in *Psychological Perspectives* 51 (1, 2008): 70–86, esp. 75–77, rather than the somewhat rougher version that is found in the Chinese conference proceedings, pp. 211–230.

11. The Chinese term had been introduced by Richard Wilhelm in *The Secret of the Golden Flower: A Chinese Book of Life* (San Diego, CA: Harcourt, Brace, 1931/1962), pp. 34–36 as referring to a contemplative practice of "reflection" during Taoist meditation, a specific technique that reverses "the self-conscious heart, which has to direct itself, fixing its contemplation on the place where a formed thought arises, that is, toward that point where the formative spirit is not yet present" and from there to start contemplating again. This, Wilhelm somewhat confusingly calls "the circulation of the light" with the light being "the contemplation" and the returning to its point source the "fixation" of the contemplation. The different understanding of *fan-chao* (pinyin: *fanzhou*) offered by Cleary [*The Secret of the Golden Flower: The Classic Chinese Book of Life* (New York: Harper Collins, 1991), pp. 73, 76, n. 11] considerably clarifies this practice, which he also locates more within a Buddhist tradition. In his book *Ordinary Mind as the Way: The Hongzhou School and the Growth of Chan Buddhism* (New York: Oxford University Press, 2007), p. 223, n. 98, Mario Proceski says that the term "refers to the mind shining back on itself."

12. I am indebted to the Sinologist Victor Mair for pointing me to this text. This is a 1929 translation by Witter Bynner and Kaing Kang-hu in E. Weinberger and O. Paz, *Nineteen Ways of Looking at Wang Wei* (Kingston, RI: Asphodel Press, 1987), p. 10.

13. C. G. Jung, *Visions: Notes on the Seminar Given in 1930–1934 by C. G. Jung*, ed. Claire Douglas (Princeton, NJ: Princeton University Press, 1997), pp. 758–59.

14. Chu His (Zhu Xi), lived 1130–1200; See Kidder Smith et al., *Sung Dynasty Uses of the I Ching* (Princeton, NJ: Princeton University Press, 1990), pp. 167–205.

15. Chuang Tzu (Zhuang Zhou or Zhuangzi), 4th century BCE.

16. John Beebe, "Achieving the Image," *Zhouyi Studies* (2009–2010): pp. 32–40, esp. 33–35.

17. Roderick Main, *Revelations of Change: Synchronicity as Spiritual Experience* (Albany, NY: State University of New York Press, 2007), p. 187.

18. Main, *Revelations of Change*, p. 186.

19. Joseph Cambray, "Towards the Feeling of Emergence," *Journal of Analytical Psychology* 51 (1, 2006): 1–20.

20. Wilhelm, *The I Ching*, p. 237.

CHAPTER 16

FROM TRADITION TO INNOVATION
WHAT HAVE WE LEARNED?

CATHERINE CROWTHER AND JAN WIENER

Our aim in this chapter is to evaluate the effectiveness of the remote training model now that we have enough distance from our own involvement. We feel we have accumulated both depth and breadth of experience and learning; depth from almost twenty involving years as project coordinators and supervisors in Russia, and breadth from Jan Wiener's engagement with many different groups as former International Association for Analytical Psychology (IAAP) Vice President and Regional Coordinator for Eastern Europe. Additionally, both of us have taught abroad and returned home with rich impressions of each place. Reflecting on our impressions and drawing some conclusions now feels timely. We have found it stimulating and thought-provoking to do this in conjunction with the eloquent accounts of the other authors of these chapters from all their disparate settings. We want to gather up some of the common themes in the book and examine the areas of similarity and difference that have emerged in its compilation.

DESCRIPTION AND HISTORY OF THE RUSSIAN REVIVAL PROJECT

Our initial involvement in Russia came as a consequence of the dismantling of the Iron Curtain. The spirit of the 1990s was characterised by curiosity and openness on both sides, together with contradictory feelings of romanticism and wariness about the "other."

We entered an "interactive field of strangeness" in which neither we nor our Russian colleagues could rely on our usual interpersonal and cultural signposts; all of us found ourselves in the domain of the operation of the self.[1] The constancy and necessity of this self-to-self involvement is the enduring lesson we have taken from our years in Russia, a living confirmation of Jung's observation that any meaningful interaction transforms both parties.

We first came to Russia to work with General Practitioners (GPs) in 1996 and were introduced to Mikhail Reshetnikov, Director of the East European Institute of Psychoanalysis in St. Petersburg. He helped to facilitate the renaissance of psychoanalysis in Russia after *perestroika*. Under Stalin, all depth psychology was forbidden and it needed a decree from President Yeltsin to reinstate it. Although people in the Soviet Union had been isolated from developments in psychoanalysis for most of the twentieth century, some study of psychoanalysis went on illegally underground, and Russians seem always to have had a hunger for this way of understanding.

Mikhail Reshetnikov was keen to have a Jungian study programme in Russia, but insisted that we do it there—Jungian analysis should develop within Russian culture. With funding from the IAAP, we first developed a two-year theoretical course between 1998 and 2000. This course took place over twelve weekends in St. Petersburg with teaching from analysts from all four London Jungian training societies. The UK-Russian partnership expanded when Russians in St. Petersburg expressed their wish to train in greater depth to become Jungian analysts. We offered a programme of clinical supervision with six UK analysts each visiting St. Petersburg from the UK four times a year to supervise clinical work in small groups and later individually.

We also decided to offer a programme of "shuttle" personal analysis where four senior UK analysts visited for a week, four or five times a year, to take on trainees for personal analysis. This was something of an experiment for two reasons. First, for those Russian analysands without English, the analysis took place with an interpreter in the room. Second, those who chose to work with a UK analyst had to adapt to a particular kind of frame where sessions were not continuous throughout the year, but rather in intensive chunks with long gaps in between.

An official IAAP Developing Group (DG) was established and we had the title of Co-Liaison Officers. With continuing financial and moral support from the IAAP, plus our own fundraising efforts, the project grew. One by one, the first trainees took their intermediate and later their final exams with external IAAP examiners, and in 2004 we welcomed our first internationally recognised analytic colleagues in Russia. In 2004 the programme was extended to Moscow by two British colleagues, Penny Pickles and Martin Stone, along with a fresh team of UK supervisors and personal analysts.

We held a Jungian summer school in Kiev in Ukraine in 2006 which brought in new participants from distant regions including neighbouring post-Soviet countries. In 2007, the Russian Society for Analytical Psychology (RSAP) was formed, and since then new groups in Krasnodar in the South of Russia, Kemerovo in Siberia, and Minsk in Belarus have begun to train as Jungian analysts, using some of the "home grown" Russian analysts whom we had trained as their personal analysts and supervisors.

We began handing over responsibility to the more experienced Russian colleagues by running a supervision course in 2009, and the formal aspects of our programme were brought to a planned close in December 2010. Since then, there have been two international conferences in St. Petersburg in 2011 and 2012 and two further supervision courses. In August 2013 the RSAP was recognised by the IAAP as a Training Society in its own right. About sixty Russian analysts have qualified through our programme and it seems likely that during the next two decades Russia will become one of the largest Jungian communities within the IAAP.

It is difficult to convey how something that started so modestly in 1998 as a seminar series has grown organically into a full analytic training, and how much we have "learned on the job." The progression of the Russian Revival Project was not foreseen by us at the outset. We all felt ourselves seizing the opportunity afforded by this significant moment of history and responding to the needs that were expressed. We have come to respect the unpredicted and emergent qualities of a wider system where the whole turns out to be much more significant than the sum of its parts. This serendipity reinforces our sense of having been co-learners with the Russians in a creative partnership where we co-constructed necessary structures and procedures where none had existed before. We also learned from clashes and disagreements. Throughout our involvement we found that we were putting things in place without fully realising the depth and enduring impact they would have. We recognise that some of the qualities of "moments of meeting" were intermittently and powerfully present, emerging spontaneously and mutually from what Stern et al. describe as the necessary long preparatory periods of "moving along."[2] The following are some key aspects of our experience.

Patient Population

As supervisors we heard about very different patient populations from those in Britain. Quoting from our 2002 paper,

> The histories of many of the patients they brought for discussion reveal the privations and harshness of childcare and family life pertaining in recent decades in Russia. Common to many histories are single parent families, babies left with *babushka*, children boarding at nurseries at a young age, abortion, alcoholism and violence. Grandparents, who play such an important role in forming early object relations, often die young because of poor health care and hard lives. In Russia the probability of dying between the ages of twenty and sixty is 45%, compared with a mere 10% in Western Europe. Quite often we hear about early diagnoses of neurological problems in children where an alternative explanation might have been one of emotional distress. The degree of trauma depends in part on the social and cultural meaning attached to it. Issues of attachment, deprivation and separation are ubiquitous, almost 'normal,' but does that mean people suffer it less?[1]

Although living conditions have since improved considerably, the gap between city and country, rich and poor has enlarged, and the sense of a country recovering from societal, cultural, and personal trauma is never far away.

Cavalli notes that when working transculturally in the process of recovery from trauma there is one universally pertinent protective factor—namely the existence of good internal objects in infancy.[3] She works in analysis with traumatised patients less in need of "reconstruction" but of "construction," to create for the first time new good internal objects and to counteract an internal deadness where nothing worthwhile exists. This concept has particular relevance to many of the Russian patients we heard about in supervision where historical and inter-generational trauma were played out at the personal level.

Adjustment and Adaptation

When we first arrived, our collective mutual projections between West and East were simple and primitive. We shared with the Russian students a joint assumption that we were the teachers and they the learners. Very quickly this became a much more complex issue, in which we had to become learners about the psyches, daily life, and professional traditions of our students, who were already mature psychotherapists. Likewise, they had to teach us how to communicate our knowledge in a relevant way for the Russian social conditions. We tried to distinguish what we held as fundamental principles in our analytic traditions and theories and what needed to be adapted to the different situation, without selling our values short.

One of the greatest challenges was of language with all its cultural nuances. Despite mutual attempts to learn each other's language, we had to rely at all times on the presence of a "third," our interpreters, without whom communication would have been impossible. We needed to use more careful scrutiny and processing of our transference and countertransference affects in tracking non-verbal clues from a culture so different from our own.

The most important realisation, which has sustained us and become a guiding principle throughout the length of our involvement in the interactive field of strangeness, is that it was not our teaching but "ourselves" that was needed, and that we had to risk sailing out on the unknown, open sea—co-creating a new space for an experience of dialogue, mutual curiosity, and a negotiated relationship. We continuously revisited the question of how to help the Russian trainees to use not their knowledge but themselves in their work. We encountered very formal didactic teaching models from Soviet times, an atmosphere rather harsh towards each other's ideas and expecting our criticism and judgement. We like to believe we gradually influenced their "internal working models" by using their interaction with us to internalise a process of discovery inside themselves.[4,5]

A Russian analyst, Vladimir Tsivinsky, has written about his first impressions of us at his initial interview.[6] He had expected a harsh interrogation and was surprised instead to have a "collegial talk":

> Such an interview was a completely new experience for me having trained in a Russian education system with 'right' and 'wrong' answers to questions. On this occasion, I felt free to express what I really thought and felt, and my first impression of British Jungians was very positive.

He also wrote about his experience in supervision groups and his adjustment to a different learning environment including his surprise that senior analysts could admit to mistakes and uncertainty:

> I learned a very important lesson: it is alright not to know something. This can be said openly and without shame or guilt. It was such a relief for me to discover that I didn't have to know everything! I became more open during supervisions and I wasn't afraid of being criticized by the group or the supervisor. I understood that not knowing is not the same as being incompetent. Moreover, not knowing is an invitation for a mutual journey into the unknown.

We believe the essential approach we were able to teach was less about formal theory or specific technique in the consulting room, but more about modelling our use of our analytic selves. Vladimir's earlier expectation of an analyst's formal, judgemental persona was echoed in our experience as IAAP examiners for the new Router Groups in Russia and other Eastern European countries. We wanted to shift their expectations from the idea of completing an academic course to one of being ready to qualify as an analyst, combining qualities of character and competence.[7] When reading routers' case studies presented for their exams, we have sometimes witnessed a disjunction between their sophisticated understanding of theory and the way they made use of it in their approaches to their patients. It seemed difficult for them to find a fluent channel between their theoretical base and their analytic intuitions, almost segregating the two as separate standpoints of understanding. It was striking, however, how in the *viva* interviews they were able to talk feelingly, thoughtfully, and insightfully—bringing the relationship between patient and router to life and giving us full confidence in their abilities as analysts.

We wondered if it is easier for routers to take in the more classical aspects of Jungian theory. Russians, with their deep immersion in the symbolic traditions of folk and fairy stories and their sensitivity to dream material, often showed an impressive recognition of the archetypal dimension of their patients' lives. Our own background as developmental Jungians brought in a different view about helping them observe and open themselves thoughtfully to the unconscious process in the transference (and especially the countertransference) with their patients, with us, and also in the psychodynamics of their relationships within their groups.

We realised they were surprised by our style of examining in which we would try to relate to them in congruence with our analytic selves. Although it was important to test the standard of their knowledge, what we were primarily seeking was their self-awareness and capacity for self-reflection as analysts and their ability to translate theory as concept into theory applied to a living relationship. Sometimes, an exam provided a turning point and stimulus for a router's development. Throughout the whole training journey it was important that as supervisors and examiners we wrote reports transparently, making detailed observations about each router's development for him or her to digest and utilise, rather than merely passing judgement. Likewise, we paid attention when we received their critical feedback and tried to digest its meaning and implications for the future.

Boundaries and Frame

One of the major areas in which at first we found ourselves disoriented when supervising involved boundaries and frame, as we were faced with very different professional traditions regarding the meaning of time, money, confidentiality, and holidays. We were startled by the flexibility with which Russian supervisees agreed to change session times at the request of their patients, rearranging "as and when" on the phone, sometimes in response to a crisis. We questioned the frequency of double sessions, until it was explained about the long arduous journeys on public transport that patients made from the high-rise housing complexes in the suburbs. Our supervisees wanted to work in more depth and this was their practical solution.

The Russian summer break when all families go to country villages stretches from mid-June to mid-September. This does not mean there is no contact between patients and therapists over this period. Patients would phone and sometimes sessions would be arranged back in the city. Often no fixed date is arranged for the first session of the autumn but an agreement made that the patient phones when they are ready to resume. Patients sometimes announce they want a pause, or they go away on long business trips. Neither patient nor analyst seemed to take up these instances as having any particularly significant meaning in the analytic relationship, nor established whether the session time remained reserved for them over a long absence or whether cancelled sessions would be charged.

Hearing about these casual, easy alterations to the frame made us uneasy. Did it mean that supervisees were depriving their patients of the opportunity to experience some hostile feelings within the negative transference? Were issues of containment, control, dependency, and autonomy being lost? Or was there a cultural context in Russia that we didn't grasp which made flexibility more useful than a stricter boundary? Challenging their practices was met with a polite "You do not understand the Russian way," and so we embarked on an exploratory discussion, showing respect for their experience and traditions. We heard that as they tried to establish their new profession of psychoanalysis, supervisees were financially dependent on their patients and feared losing them if they did not accommodate their demands as "customers," rather than set boundaries for them as patients. Paradoxically, for Russian therapists, living with this unpredictability of income and uncertainty of time schedule seemed to offer them more security in terms of retaining their patients, and therefore maintaining their income.

Early on, referrals came from family members and acquaintances, tangling routers up in issues of confidentiality and secrecy. Our suggestion that a referral network would help preserve boundaries was a new and welcome idea, as they had not yet enough experience of resisting the immediate demands of people in distress. We began questioning supervisees' own feelings about their private lives being intruded upon by the frequency of phone calls outside session times. This often elicited relief as they acknowledged countertransference feelings of resentment, fear, or avoidance of patients' demands.

We found we had to re-examine and explain to ourselves, as well as them, the special construction we give the frame in analytic work as a container for the evolution of the transference. Our exchange of ideas helped pinpoint how, without a robust frame in place, analyst and patient are both deprived of necessary engagement with the negative transference. If there is anything we have left behind in Russia it is a respect for boundaries!

Our supervision was helped immeasurably by the new experience they were simultaneously having of their own boundaried personal analysis from the British shuttle analysts. We witnessed their work gradually becoming more empathic and reflective. This was a welcome encouragement of our controversial model of providing shuttle analysis. With all its shortcomings it provided our Russian supervisees with a relationship in which consistency and predictability were valued and feelings about absence and deprivation acknowledged and discussed.

Attachment and Reverie

The insights we brought about attachment and early infancy in transference work with adult patients resonated for many Russian routers. Natalia Alexandrova describes the "Soviet infant" as being treated in a purely functional and medical way, so that the newborn baby only saw his or her mother at strictly scheduled feeding times, five times a day, twenty minutes at a time.[8]

> At the same time the mother was expected to continue her work, and the whole family's daily schedule was adjusted accordingly. Factories and offices introduced feeding breaks, and there was an initiative of 'a special maternal conveyor' (a staggered schedule for breastfeeding mothers working in the same factory) to avoid any interruptions at work. Later that phenomenon would be called the 'double burden,' because till the very end of the Soviet regime an ideal woman, according to the national ideology, was a woman who bore children, ran the house and still worked full-time outside the home.

Alexandrova then comments on the lack of space for the mother's reverie under the strictly time-limited conditions of running back and forth between factory and baby, and makes a telling comparison with her experience of the shuttle supervisors and trainers as breastfeeding working mothers:

> It's a paradox: your feeding time rigorously complies with the schedule (twelve visits of supervisors and four visits of analysts per year), but you still have no certainty. ... Our supervisors and analysts—like our mothers before them—existed in two spaces: their work in London and their shuttle visits to Russia. Like our mothers, they had both to work and to 'breastfeed'; time was scarce; they felt tired and sleep-deprived. Like our strong and vigorous mothers, our supervisors and analysts had to get up at five in the morning to meet the 'feeding' schedule; they flew into a different time zone to have their working day start three hours earlier. And then went back to their London work till the next 'feeding break.'

Alexandrova's insightful reverie about our crowded schedule had a real impact on our awareness that the shuttle system did not allow for wasting time creatively and sometimes adversely affected our ability to nurture and think about our routers.[9] It was only on the plane home after a busy weekend that we had the space to process our feelings. The powerful transferences onto us as containers for the whole project reinforced again that what they needed was *us*. How could the shuttle model ever sustain a "good enough" relationship of containment and maternal thoughtfulness? We believe that despite their experience of being picked up and dropped, our constancy and endurance over the length of nearly twenty years' commitment did make a difference.

Appraisal of Shuttle Training

We were concerned how to maintain a coherent and containing learning environment with such infrequent visits when, by contrast, we ourselves were used to the privilege of regular analysis and supervision for our work. Issues of attachment and separation were paramount and felt most profoundly at the beginning and end of the programme. Although we tried to offer some containment and predictability by setting a regular rhythm for visits and planning the schedule long in advance, infrequent visits made sustaining a "good enough" quality of relationship complicated. The shuttle programme mirrored transgenerational patterns of losses experienced by so many of our Russian colleagues. We learned that personal analysis was sometimes unsustainable for those trainees with either borderline or schizoid features, even with Skype sessions in between. For trainees with less access to their feelings, or weaker ego function, the shut-down between visits made progress slow. There were a small number of people who dropped out from training for these reasons; however, for the majority, there was something valuable gained from the shuttle experience, despite all the obvious and very real shortcomings. There is recognition too that shuttle programmes are not for everyone as they demand high levels of ego strength during absences and patience and determination to last the course.

One unforeseen consequence was their identification with our shuttle model of work. We noticed that many of the Russian analysts immediately after qualifying began to offer shuttle programmes to other regions both inside and outside Russia which inevitably left their patients at home dealing with separations and feeling neglected. This reproduction of a shuttle model probably should not have surprised us, yet our expectation had been that while this first generation was finding a way to train, given the lack of analysts in their country, the second generation would be offered a steady, consistent training in their own cities. We hope that the establishment of the RSAP training in 2015 will establish a home base for analysts, trainees, and analysands alike.

Pivotal Events

The history of our project has included some key serendipitous moments that arose organically from the needs of the situation and between our two cultures, and which have had lasting effects. In 2006 the idea for a Jungian summer school emerged spontaneously and was organized by our Ukrainian colleagues in Kiev. With the attendance of some senior IAAP officers and an unprecedented participation from new individuals and groups from far-flung regions of Russia and neighbouring countries, it was a turning point in acknowledging that a movement for Jungian training was underway in post-Soviet space. Collaborations with foreign colleagues developed, and a sense of being part of a worldwide Jungian community was articulated. From Kiev 2006 onwards, outreach teaching in distant parts of the Russian

Federation (which Russian analysts from Moscow had long undertaken) would accelerate with spontaneous arrangements for inter-regional learning, supervision, and analysis.

The most recent pivotal event occurred in 2014 when the Serbia Developing Group hosted a supervision course open to all analysts in post-Soviet space. Because of perceived personal tensions, lack of privacy, and gossip in their small Societies, participants from several countries were relieved to share their work with foreign colleagues in a more anonymous and confidential space—this time amongst people with shared histories like themselves. We have heard that after the course significant new working partnerships have emerged quite spontaneously through Skype and face-to-face meetings. Significantly, requests for more interactive "master-class" learning provided valuable feedback which will undoubtedly influence the planning of future courses.

In both these examples we see a de-integration process creating autonomous life beyond our predictions with newly qualified analysts shaping their own regional community of Jungians.[10] These pivotal encounters are analogous to mutual "moments of meeting," creating interpersonal dynamics and planting the seeds of new partnerships.[2]

Leadership, Authority, and Organisational Dynamics

Embarking on work in another culture—often one with a radically different social history—results in working partnerships where external and internal experiences of authority figures are likely to affect the negotiation of "good enough" working relationships. Clashes of cultural identities can make for difficult group dynamics. For those of us brought up in a culture of democracy, a system of involvement of the people in political decision-making is taken for granted. In the East, *perestroika* brought with it the potential to value individuality, but as Solzhenitsyn succinctly says, "in seventy-year-old totalitarian soil, what democracy can sprout overnight?"[11] Sebek emphasises the inevitable time-lag and persistence of internal totalitarian objects in the psyche even when the external totalitarian power seems to have disappeared.[12] Because our social and political histories are all different, cultural complexes cannot but emerge in cross-cultural partnerships around issues of authority at personal, group, and organizational levels.

We are told in the Western media that Russians have always had a passive attitude to authority. Anna Politkovskaya—anything but a passive voice—frequently goaded her fellow Russians to notice the erosion of their hard-won freedom of speech.[13] She identified the Duma's lack of parliamentary debate and railed against the Russian people declining to stand up to the rise of so-called "Putinism," that is, the gradual centralising of control in the Kremlin, which we have watched during our time in Russia. Many of our Russian colleagues feel disbelieving about the possibility of an effective political opposition, which is hampered by the absence of three fundamentals: an independent judiciary, independent mass media, and independent sources of finance. We were told how in Soviet times, if authoritarian control was not passively submitted to, this would lead to defiance, evasion, using trickery, or "peasant cunning."

Our official roles in Russia on behalf of the IAAP were as co-liaison officers and supervisors. The word liaison is not itself without mercurial overtones—dangerous liaisons. In cookery, the term denotes a thickening for sauces, mixing different ingredients together smoothly. Disputes around issues of authority were often the lumps in the sauce that threatened to ruin the dish!

Giving authority to a leader (in our cases our appointment by the IAAP as co-liaison persons for St. Petersburg) implies the need for followers. As Obholzer remarks, "leadership would be easy to achieve and manage if it weren't for the uncomfortable reality that without followership, there could be no leadership except perhaps, of a delusional sort."[14] For the first few years of our work in Russia our roles as leaders were met with tacit agreement to follow. As our relationships evolved, so the distinctions between leaders and followers gradually (and healthily) became less sharply defined, and a more collegial and symmetrical structure developed. Occasionally, there was less collegiality and the clashes that erupted often happened when understanding broke down as a result of mutual projections where authority was usurped by a perceived authoritarianism.

Analysts are interested in the complex relationship between internal states of mind and professional, group, and organizational functioning, and exploring ways in which unconscious processes can affect the successful implementation of individual and group tasks. This is an approach that Huffington et al. have called "working below the surface," drawing on a subtle combination of psychoanalytic theory, systems thinking, and group dynamics.[15] They make a helpful distinction between "hierarchical leadership" and a much flatter hierarchy they call "distributed leadership." From hierarchical beginnings where we were welcomed because of our expertise and because we were perceived to have something of value to offer, we had to try to evolve a more distributed style of leadership involving devolution of decision-making, increased collaboration, and the gradual passing of authority and autonomy to our Russian colleagues. This shift was sometimes difficult, removing from us some of the inevitable safety, narcissistic gratifications, and inflations that go with a more hierarchical structure.

Obholzer considers that leadership involves vision, strategy, management of boundaries and change, and containment.[14] While there was a vision in our partnership with routers in Russia and a strategy for achieving it—to train Jungian analysts and form new IAAP Societies—working towards mutually agreed strategies for change in a contained way with only irregular visits was never easy. Obholzer talks of

> managing a quite sensitive titration process where too many external ideas overwhelm in-house values and the strengths of the past are lost; too little titration of external reality leaves the organization at risk of being bogged down and isolated. Titration involves having to hold a number of different issues in mind.

This proved a key difficulty for us in moving from hierarchical to distributed styles of leadership. The following are a few examples of "the lumps in the sauce."

Teaching Styles

The most striking memory from the earliest stage of our work was the shock of being expected to conduct a seminar in a huge auditorium with ourselves lecturing from behind a desk on stage while the audience in fixed tiered seats would stand to address questions to us. This embodied the Russian expectation of hierarchical leadership. There was neither dialogue nor exchange between the audience members, and we were all communicating through the single interpreter. By the end of the day we had diffused this formal enactment of our expert authority, decamping to a crowded smaller room and using exercises in which participants talked in small groups. The Russian tradition of learning is didactic and top-down with all the authority invested in the teacher. There is naturally always a shadow of scepticism and ambivalence about this arrangement which remains hidden and therefore subversively powerful.

Claiming Authority

A new constitution for the Developing Group in St. Petersburg contained elected roles such as Chair, Honorary Secretary, and Treasurer. Problems began to arise when those in positions of authority found themselves the recipients of powerful and often negative projections especially when IAAP subscriptions were due and DG members showed some reluctance to part with their money. At times disagreements could not be resolved by debate or democratic votes, and, instead, leadership and decision-making were seen through the prism of dictatorship and victims. The leaders struggled to win the consent of the followers and retain their authority. Delegation was difficult, with some members of the group reluctant to join in and share out the tasks of the DG. The attitudes we tried to import about working and learning in groups and professional communities, and our encouragement of a common purpose for the Developing Group, came up against the difficulty for them to take ownership of their own structures and processes. We learned that they were not meeting as a group unless members of our team from London were present. Most of the routers were used to strong Russian leaders, not only their politicians, but also in the Institute in which many had completed their first psychology training. They had no previous experience of any organisation with a purpose, constitution, elections, dues, membership rules, or codes of ethics, except of course the Communist Party, whose very membership denoted strict adherence to the Party line. Individuals willing to take on roles could become scapegoated and personally wounded that their efforts were met with criticism. It felt at times that the complexity of the web of projections and unconscious expectations around issues of authority versus autonomy would not be containable.

Transitions towards Distributed Leadership

A culture clash played itself out painfully around the formal process of moving from hierarchical to distributed authority in the formation of the new RSAP. Our pride that the number of Russian analysts had now reached the required critical mass to apply for recognition by the IAAP probably led us to push for this to happen too quickly when the analysts were not ready, and there was much internal and external planning needed between sibling analysts in Moscow and St. Petersburg to work out their own vision and strategy for their future. We became caught up with the IAAP's eagerness for new Societies. With hindsight, we realise we had not built in sufficient consolidation time after qualification for the Russian analysts to work through their own ambivalence, rivalries, and anxiety to discover their own organisational ethos and independence. Because there was no existing institutional infrastructure to receive and contain them after they graduated, there was a residual dependence on us. We were by now dividing our time between them and the new generation of routers who had just joined the programme, taking the places the analysts had vacated. We were asking them to "leave home" and go out into the world too abruptly, without sufficient mentoring or consultation, repeating unconsciously so many previous ruptures and enforced separations Russians have had to endure. The question—whose vision and strategy, ours or theirs?—needed to be addressed with our Russian colleagues on many occasions.

Some Common Themes

We read with interest the chapters from the other authors, keen to observe any common themes and differences. The most striking commonality amongst the authors is that everyone has found his or her intercultural encounters moving, challenging, meaningful, and life-changing. Most people have written very personally about the resonances with their own inner lives. Authors from both sides of training write of their long-term commitment to their specific intercultural partnerships, a loyalty that attests to the high value placed on the various projects by everyone involved. The sense of a "calling" to depth psychology and the inner readiness for a developmental journey is common to both trainers and trainees.

A related theme concerns the quality and nature of the knowledge transmitted through training. All authors emphasize the vital ingredient of the teachers showing a personal emotional engagement that exemplifies and models an embodied analytic attitude even in the formality of theory teaching. Richard Sennett, the sociologist and philosopher, writes about craftsmanship as "an enduring, basic human impulse, the desire to do a job well for its own sake."[16] He emphasizes how complicated craftsmanship can be because of the conflict about standards of excellence affected by peer pressure, obsession, or frustration. Sennett highlights how the craftsman's essence is "the special human condition of being engaged."

Sennett makes a very helpful distinction between "embedded knowledge" and "explicit knowledge." Embedded knowledge is learned from good teaching and with plenty of practice becomes embedded as a set of skills. By contrast, explicit knowledge implies self-awareness about what we do and know. It brings with it a need to put this knowledge into words for others, to ensure the transfer of knowledge.

All the authors have brought their valuable embedded knowledge to contribute to the various intercultural encounters. Translating some of it as explicit knowledge is the manifest purpose of this book. But what is vividly described time and again in each chapter is that conveying the "real" message occurs in the actual processes between people, whereby embedded knowledge is lived out and exemplified in face-to-face engagement. Sometimes, as teachers, we have conceived ourselves as architects and designers to make the foundations safe for the joint project. It is for the students to develop the work and build the structures that will endure.

The next common theme evinced in all the chapters shows the bi-directionality of influence (as Joe Cambray names it) in partnerships. It may sound disingenuous to refer to any training project as a partnership when there is an evident distinction between the teachers and the students. Usually, at the outset of a project—like the start of an analysis—it is an asymmetrical partnership, as Ogden describes.[17] Yet one of the powerful messages of this book is that uniquely in this field of intercultural exchange learning takes place on many different levels within each one of the participants. Just as the analyst learns anew from each patient, so the teachers have to open themselves to unfamiliar experiences and rely on the students to induct them into the values of a different culture and teach them how to teach. More tellingly, both sides have to find a capacity to tolerate "not knowing beforehand."[18] The evidence of these chapters is that growth and mutual revelation occur in all participants regardless of status or length of experience, and that partnership is a goal that can be aspired to and achieved by both sides.

Another common theme is how the encounter with the "other" allows the shadow to become explicit and be met. It concerns an acknowledgement that our unconscious ethnocentricity attends every encounter, with assumptions, projections, and stereotyping of the "other," bringing with them the responsibility to monitor our preconceptions and guard against acting out our cultural shadows. One way in which the authors have reflected on this is by studying the history, politics, and culture and in some cases the language of the other nationality. One cannot work effectively without knowing something of the culture in which one is working, and this requires continuous mutual comparison between the participants. Intergenerational cultural trauma is frequently mentioned. It seems that the recounting and describing of cultural and historical events has a healing effect when both parties can use the narrative to give shared meaning to the present situation and release it from the shadow affects of shame, envy, competition, guilt, and anger. Related to this, several authors have touched on the interplay of unconscious projections concerning the dynamics of power and authority in training.

The use of language and the "mother tongue" is another important theme mentioned by all the authors. The capacity of translated language to both communicate and obscure meaning is subtly acknowledged and celebrated with many examples. But although there are frustrations to be weathered in negotiating a different language, no one has written that in his or her experience it made for any less rich emotional contact. There are also rewards in slowing down the process and allowing oneself to become more attuned to the nuances and specific cultural expressions of the other language, which reveal imaginally different habits of mind. The complexity and responsibility of the task of the tireless interpreters deserves and receives full tribute.

DIFFERENCES OF VIEW

Some themes have evoked disagreement. For example, there is no settled view about the shuttle system of training especially where personal analysis is concerned. The ethical question is about balancing the damage caused to the analysand by all the separations against the necessity to offer what is possible by way of a recognised IAAP analysis in order to get something started in a new location. The persisting requests from Developing Groups for established IAAP analysts to come as shuttle analysts sometimes seem to denigrate what the recently qualified local analysts can offer in terms of cultural identification, mother tongue, consistency, and steadiness. The capacity of most patients to form a transference attachment to a shuttle analyst does not seem to be in much doubt, but how well the relationship can contain and digest negative feelings about loss over exiguous time and space has a different fate depending on the pre-existing ego strength of each patient.

Another disputed theme concerns the limitations and benefits of being the outsider. The worries about missionary zeal, imperialism, or colonising are frequently mentioned. But, on the other hand, for some traditional cultures there can be freedom from constraint in talking confidentially to an outsider who is not bound by the same social norms. Different perspectives can be welcomed or suspected. As Sennett puts it, "For the foreigner, the knowledge that he comes from elsewhere, rather than being a source of shame, should be a cautionary knowledge."[19]

Similarly, despite the apparent success of so many training schemes around the world, the capacity of the IAAP to give firm shape and a containing boundary to a shuttle programme is questioned. Does a group history of shuttle training affect the subsequent ethos and infrastructure of new local training institutes, or, in other words, is deprivation unconsciously built in to their expectations? What comes across in several of the accounts of shuttle training models is that in the absence of a clear containing structure or in periods of disagreement at an institutional level it is personal relationships with colleagues both within the group and with international visitors that come to be relied on in compensation. Issues of trust, leadership, authority, hierarchy, and power have to be hammered out between individuals. It is hard to establish any agreed values within a new Society when there are many different traditions and principles at stake and where there is no pre-existing infrastructure beyond the remote-seeming IAAP.

Conclusion

In writing this chapter, we have held in mind three overlapping but different aims: First, to convey to the reader something of our learning from working in Russia (mainly St. Petersburg) for nearly twenty years, collaborating with our Russian colleagues as participants in their professional development from their first beginnings to their emergence as a new IAAP Society. Secondly, we have looked at whether our experiences echo the learning and reflections of other authors in our book—the teachers, analysts, supervisors, and those who have trained through the IAAP Router Programme. Finally, we gather up the threads of our experience to take us forward to more effective collaborations.

The interactive field of strangeness constellated in cross-cultural partnerships is one in which both partners become deeply involved. In this sense, the work honours Jung's conviction that transformation can only occur if both parties are involved and changed. As Schore puts it,

> Intersubjectivity is thus more than a match or communication of explicit cognitions. The intersubjective field co-constructed by two individuals includes not just two minds but two bodies. ... At the psychobiological core of the intersubjective field is the attachment bond of emotional communication and interactive regulation.[20]

The most important finding from our own work and that of others is that effective learning comes from a developing ability to trust in the activity and actions of the self. We show ourselves as the clinicians and human beings that we are to those whom we help to train. It is this that seems to have given our trainees a capacity to internalise how they may use themselves as developing analysts (in their own cultures in their individual ways with their own patient populations) and to reward us with their fresh insights.

Have these programmes been successful? We imagine the reader will agree from the authors' and our own contributions that they have indeed been valuable, and we now have an analytic culture that is blossoming in new parts of the world. In the past it was theory that influenced practice, but working in different cultural settings permits practice to make a greater impact on theory—a topic perhaps for another book?

The IAAP's remote training programmes offer guidelines for structures in which Developing Groups may grow (liaison officers, router coordinators, visiting teachers, supervisors, personal analysts, examiners, etc.) as well as guidelines for what routers need to do if they wish to qualify as internationally-recognised Jungian analysts. There is now in progress further consideration of continuing professional development for analysts in regions where first generations begin to train others, including courses to train new supervisors; however, there is an urgent need for further research that can evaluate different models of learning based on routers' experiences. It is clear that unless there is a firm container/setting, and visits are sufficiently frequent and stable, routers do not feel sufficiently well-held during their training. This is true whether we are talking about personal analysis, supervision, or teaching. While personal motivation is clearly a guiding force in any training, and people find what they can to use productively, learning is best done within a safe and holding structure where there is space and time, mutually, to gather up the process and reflect on what works and what does not.

We end our chapter with a quote from Albert Camus relevant to our own learning: "One may long as I do for a gentler flame, a respite, a pause for musing. But perhaps there is no other peace for the artist than what he finds in the heat of combat … every wall is a door."[21] Negotiating the walls to find new insights has indeed fundamentally challenged all we know from years practising our craft of analysis. We have discovered that learning and using imagination creatively can emerge from the experiences of difference, primarily in the establishment of lasting and deep personal connections with people from another culture.

Notes

1. Catherine Crowther and Jan Wiener, "Finding the Space between East and West: The Emotional Impact of Teaching in St. Petersburg," *Journal of Analytical Psychology* 47 (2, 2002): 285–301.

2. Daniel Stern et al., "Non-Interpretive Mechanisms in Psychoanalytic Therapy: The 'Something More' Than Interpretation," *International Journal of Psychoanalysis* 79 (1998): 903–21.

3. Alessandra Cavalli, "Transgenerational Transmission of Indigestible Facts: From Trauma, Deadly Ghosts and Mental Voids to Meaning-Making Interpretations," *Journal of Analytical Psychology* 57 (2012): § 597.

4. John Bowlby, *Attachment*, vol. 1, *Attachment and Loss* (London: Hogarth Press, 1969).

5. Jean Knox, *Archetype, Attachment, Analysis: Jungian Psychology and the Emergent Mind* (Hove: Brunner-Routledge, 2003).

6. Vladimir Tsivinsky, "Impressions and Reflections," in Catherine Crowther et al., "Fifteen Minute Stories about Training," *Journal of Analytical Psychology* 56 (5, 2011): 635–36.

7. Jan Wiener, "Evaluating Progress in Training: Character or Competence," *Journal of Analytical Psychology* 52 (2, 2007): 171–85.

8. Natalia Alexandrova, "The Double Burden of Shuttle Analysis," in Crowther et al., "Fifteen Minute Stories about Training," pp. 642–43.

9. Thomas Ogden, "On Psychoanalytic Supervision," *International Journal of Psychoanalysis* 86 (2005): 1265–80.

10. Michael Fordham, "On Not Knowing Beforehand," *Journal of Analytical Psychology* 38 (2, 2003): 127–37.

11. Alexander Solzhenitsyn, *The Russian Question* (London: The Harvill Press, 1994), p. 96.

12. Michael Sebek, "The Fate of the Totalitarian Object," *International Forum of Psychoanalysis* 5 (4, 1996): 289–94.

13. Anna Politkovskaya, *A Russian Diary* (London: Harvill Secke, 2007).

14. Anton Obholzer and Vega Zagier Roberts, eds., *The Unconscious at Work: Individual and Organizational Stress in the Human Services* (Hove and New York: Routledge, 1994).

15. Clare Huffington, David Armstrong, William Halton, Linda Hoyle, and Jane Pooley, eds., *Working Below the Surface: The Emotional Life of Contemporary Organizations*, Tavistock Clinic Series (London: Karnac, 2004).

16. Richard Sennett, *The Craftsman* (London: Penguin/Allen Lane, 2009).

17. Thomas Ogden, "Reconsidering Three Aspects of Psychoanalytic Technique," *International Journal of Psychoanalysis* 77 (5, 1996): 897.

18. Michael Fordham, "Integration-deintegration in Infancy," in *Explorations into the Self, Library of Analytical Psychology*, vol. 7 (London: Academic Press, 2005).

19. Richard Sennett, *The Foreigner* (London: Notting Hill Editions, 2011), p. 80.

20. Allan Schore, *The Science of the Art of Psychotherapy* (New York: W. W. Norton & Co., 2012), p. 80.

21. Albert Camus, *Resistance, Rebellion, and Death* (New York: Vintage Books, 1960).

Index

A

Abramovich, Roman, 81
Abramovitch, Henry, 53–69
Absence from analysts, 211–212
Adams, James Truslow, 223–224
Adams, Michael, 72
Akhmatova, Anna, 205, 209
Alder, Stephan, 91–109
Alexandrova, Natalia, 134–141, 211, 281–282
Allemann, Fritz, 264–265
Analysands: abandonment by shuttle analysts, 204–205; acknowledging cultural context of, 61; communicating with via Skype, 267–268; conversing via translators, 163–166; having analysis in foreign languages, 172–174; projections onto interpreters, 201–202; rage at shuttle analysis, 211–213; reaction to trainers by, 46, 205–206; recognizing vulnerability of analyst, 261–262; sharing silence and speech in training, 213–215; using sandplay, 234
Analysts: communicating via Skype, 267–268; confronting superego of, 57–58; cultural biases and countertransference by, 8–9; disorientation in Russia, 44–47; empathizing with host culture, 30; expressing vulnerability to analysands, 261–262; individuation of trainee, 58–59; infected with patient's illness, 187; leading groups, 101–103; Osterman's approach as, 258–259; as participant, 187; positioning patients in class hierarchy, 85–86; process of analysis in foreign languages, 172–174; response of routers to visiting, 59; reverie needed by, 282; seeking authority as, 44; seen as other, 222; self-assessing capacities and limitations, 208; sharing silence and speech in training, 213–215; transformative process in training Taiwanese, 235; value of intercultural training encounters to, 288; working with cultural differences, 5–6
Analytical psychology. *See also* Training: accommodating cultural differences in, 1–6; cross-cultural bi-directionality in, 24–29; culturally redefining supervision for, 8–9; developing empathy for cultural diversity, 26–28; dissemination of, 24–25; effect of Eastern cultural influences on Jung, 21–24; enforcing behavioral and political norms, 12; initiatory process of training in, 146, 148–149; introduction in Lithuania, 12; recognizing author of process in, 123–125; role of language in, 174–175; sameness between participants in, 115; Self and

individuation central in, 37; spread of, 53–54; working with words that cannot be translated, 61–64
Anamorphose, 169
Anzieu, Didier, 78
Apartheid, 83
Archetypal patterns. *See also* Individuation; Self: arising in collective unconscious, 4–5; couple in, 13; cultural commonality of, 4–5; cultural influences on Mother and Father, 6–10; definition of, 266–267; individuation and cultural diversity, 10–13; individuation as key, 37; morality and cultural shadow, 75–77; Self in, 15, 37; shadow awareness and integration, 13–14; underlying human behavior, 3; understanding cultural coding of, 61
Archetype (Stevens), 76
Asia. *See also* China; Eastern cultures; Taiwan: Chinese/Taiwanese cultural relationship norms, 226–227; Confucian principles of behavior, 190, 193; face and collective harmony in, 188; training in, 9–10, 185–187
Assumptions, 2, 4
Astor, James, 144
Attachment issues, 281–282
Authority: autonomy vs., 287; balancing in training models, 183, 184; cultural influences on, 44; issues emerging in shuttle analysis, 284–285; represented in trainee dreams, 120–123; response of post-Soviet routers to visiting analysts, 59; scapegoating of those in, 287; trainer and student issues of power and, 44–45

B

Balint, Michael, 94
Bauman, Zygmunt, 71, 73–74
Baynes, H. G., 72
Beebe, Beatrice, 20
Beebe, John, 44, 255–271
Beeghly, Marjorie, 20
Behavior: bi-directional influences on, 20; Confucian principles of, 190; psychiatry for enforcing norms of, 12
Benedict, Ruth, 188
Benghozi, Pierre, 169
Bi-directionality: concept of, 19–20; exploring in cross-cultural analytical work, 24–29; multicultural insights into Jung's work with, 30–31; noticed in intercultural exchanges, 289–290; seen in environmental models, 25–26; shifting from knowledge to interactive learning, 32–33
Bick, Esther, 78
Bion, Wilfred, 239
Blake, William, 221
Bollas, Christopher, 186, 191
Bonder, Bette, 20
Borderland experiences, 153–158
Bortuleva, Elena, 66, 82
Boundaries: correcting cultural biases by using, 8–9; created with apartheid, 83; developing in groups, 96–97; encountering differences in, 45–46, 202, 206, 279–281; experience of skin as, 78; individuation and, 46; shadows and, 77; skin and ego, 77–79
Bright, George, 181–182, 192–193
Brodsky, Joseph, 212, 215
Bruzzone, M., 184
Bryusov, Valery, 127
Buddha, 258, 264

INDEX

C

Cambray, Joseph, 19–36, 267, 289
Camus, Albert, 293
Carter, Linda, 20
Carvallo, Eduardo, 31
Casement, Patrick, 184
Cavalli, Alessandra, 237–253, 276–277
C. G. Jung Institute of Zürich, 183
Chaing Kai-shek, 229, 255–256
Chang, Jenny, 185
Chatwin, Bruce, 77
Chianese, Domenico, 193–194
Children: creating Special Time for, 239–241; developing minds of own, 237–238, 247; effect of educators holding them in mind, 244–245; infant research on ego boundaries and skin, 77–79; institution as family for, 252; not being in anybody's mind, 247; repetition of trauma enacted by, 250–251; running away of street, 241–244, 249, 250; survival anxieties of, 247
China: Beebe's address to Jungian conference in, 257–262; Chinese/Taiwanese cultural relationship norms, 226–227; Cultural Revolution in, 228–229; Developing Groups in, 185; family practices in, 224–226; finding cultural dichotomies with American culture, 221; introducing psychotherapy to, 256–257
China on the Mind (Bollas), 186
"Chinese Modernity and the Way of Return" (Liuh), 30
Chrysanthemum and the Sword (Benedict), 188
Chu Hsi, 265
Chuang Tzu (Zhuang Zhou or Zhuangzi), 265
"Civilization in Transition" (Jung), 71
Clark, Giles, 215
Clarke, J. J., 23, 257
Class structure in Britain, 84–86
Coleman, Warren, 192, 212, 213
Collected Works (Psychology and Religion: West and East) (Jung), 21
Collective: fueled by collective unconscious, 39–40; relationship between individual and, 38–39; traumas of cultural brutalities to, 39, 47–49
Collective harmony: Chinese emphasis on, 223–224; conflicts with individualistic point of view and, 225–226; Jung's thoughts on, 188, 190–191
Collective unconscious: effect of trauma to feminine, 131–133; intertwining of political and personal in, 85; Jung's archetypes of, 4–5; relationship between culture and individual in, 73
Colman, Warren, 181
Colonialism: cultural norms accepted during, 4; trainers participating in training, 55–56; traumas to individuals and collective in, 39, 47–49
Complexes, 74
Confession, 58
Confucian tradition: behavioral ethics in, 190; development of Asian women and, 193
Confucius, 258
Coniunctio: defined, 219; reaching point of, 227
Connolly, Angela M., 46–47, 135, 179–197
Containers. *See also* Boundaries: concept of containment, 237; created by interpreter and trainer,

201–202; cultural, 136; groups working as transformational, 99–100; training without institutional containment, 146–147

Cortés-Iniesta, Sandra, 252

Countertransference: cultural aspects of, 26–27; cultural-specific relationships shaping, 63, 118; with patients, 8–9; with routers, 60; using musical, 204–205

Couple as archetypal pattern, 13

Crowther, Catherine, 44, 45, 46, 47, 57–58, 273–295

Cultural complexes: archetypal morality and cultural shadow, 75–77; Baynes' concept of cultural psyche, 72; concept of, 74–75; conjunction of Chinese and Jewish cultural traditions, 229–230; discovering Russian, 79–82, 98–100; effect on organizations, 92–93; encountered in Kazakhstan, 95–98; gender roles and development of groups, 96–98; group dynamics and partnerships in, 91–109; Jewish, 229–230; Jung's appreciation of primitive cultures, 4; maturing beyond, 24–29; *mianzi,* 188–189; noticing as analyst in new environment, 92; personal complexes linked to, 83–84, 116–118; reaction to analysts by analysands, 46, 205–206; receiving supervisees in context of their, 61; rooted in cultural unconscious, 41–42, 72, 74, 192–193; search of identity shaped by, 118–120; South African, 82–84; superiority and inferiority demonstrated by, 42–44; Switzerland, 101–107; totalitarianism and Russian, 200–201, 284–285; training in different cultural contexts, 187–188; transformations in groups when freed of, 152; uncovered in Krasnodar, 98–100; United Kingdom, 84–86; Western vs. Eastern, 219–230

Cultural containers, 136

Cultural diversity. *See also* Cultural complexes; Cultural identity: absence of individuation in many cultures, 10–11; adjusting and accepting, 49; Eastern influences on Jung's ideas, 21–24; empathy in cross-cultural analytical work for, 26–28; encountering when teaching analytical psychology, 1–2; enforcing behavioral and political norms with psychiatry, 12; explaining within Jungian theory, 23–24; finding mutual understanding in midst of, 220–221; impact on student supervision, 8–9; importance of cultural knowledge in clinical discussions, 26, 28–29; influencing Mother and Father archetypes, 6–10; introduced in early analytical psychology, 24–25; sources of, 2–3; words that cannot be translated, 61–64

Cultural identity: developing for Lithuanians after Soviet regime, 111–112; factors shaping, 118–120; fear of losing, 23; of supervisors, 100, 101

Cultural Revolution in China, 228–229

Cultural shadow, 75–77

Cultural unconscious: American vs. Chinese Dream in, 223–224;

INDEX

merging collective and personal influences in, 59, 72; root of cultural complex in, 41–42, 72, 74, 192–193
Cwik, August, 124

D
Daoism, 191
Dawkins, Richard, 73
Deintegration, 78
Developing Groups of IAAP: in China, 185; Jung's possible views of, 67–68; origin of, 55–56, 274–275
Dhar Prabhu, Kusum, 39
Dialogical writing of Jung, 191
Ding, 265–267
Disciple's dilemma, 57–59
Distributed leadership, 286, 287–288
"Do Be My Enemy for Friendship Sake" (Gordon-Montagnon), 77
Doctor-patient relationships, 94
Dolto, 134
Double deprivation, 247
Dreams: authority figures in trainee, 120–123; effect on ego during, 263–264; images of death in, 139; turning light of consciousness in, 264
DSM-III, 12
D'Souza, Anjali, 39
Duffell, Nick, 85

E
East European Institute of Psychoanalysis, 202, 274
Eastern cultures: approaches to unconscious, images, and the psychoid, 192–193; comparing with Western cultural complex, 219–230; individuation in Western and, 189–191; Jungian theory and states of mind in, 191–192; Western analyst as other in, 222
Echo, 165
Ego: center of consciousness in West, 192; derived from bodily sensations, 78; effect of dreaming on, 263–264; skin and boundaries of, 77–79
Eitingon, Max, 182, 183
Embedded vs. explicit knowledge, 289
Emotions: developing cross-cultural empathic, 27–28; fear of losing cultural identity, 23; felt by interpreters, 166–168; mind unable to hide, 260–261; rage at impact of shuttle analysis, 211–213; shame associated with group, 100
Endings, 65–66
"Expressive Sandwork" (Zoja and Carvallo), 31–32

F
Face, 188
Father archetypes, 6–12
Faydysh-Krandievsky, Andrey, 211
Feldman, Brian, 78, 136
Ferro, Antonino, 181
Figes, Orlando, 72
Foden, Ann, 209–218
Fontana, Andreina, 193–194
Fordham, Michael, 78, 186–187, 239
Foulkes, Siegmund Heinrich, 93–94, 105, 107
Frame of analytic work, 280–281
Freire, Paulo, 238, 239
Freud, S., 76, 78, 80, 86, 183, 186, 191
Fu, 263
Fundación JUCONI, 238–239, 241, 242, 245, 248, 250, 252

G

Gagarin, Yuri, 211
Gao Lan, 31
Geertz, Clifford, 193
Gender roles: group development and, 96–98; Russian view of women, 128–129; women's authority as leaders, 44
German cultural complex, 100
German Society for Analytical Psychology (DGAP), 94
Gerson, Jacqueline, 168
Giegerich, Wolfgang, 40
Gladstone, W. E., 43
Golden Flower, 22, 30, 220, 257, 263, 264
Good Great Mother archetype, 83
Gordon, Elizabeth, 163
Gordon-Montagnon, Rosemary, 77
Great Mother, The (Neumann), 6
Groups: as author in analytical process, 123–125; feeling shame associated with, 100; need to belong to, 75; stabilized by interpreter, 168; stages in leading, 101–102; transference of patient and case into, 103–107; unconscious process of development in, 96–98; working as container for transformation, 99–100; working with IAAP Router Programme, 98–100
Gudaitė, Gražina, 111–126

H

Hao-Wei Wang, 185
Henderson, Joseph, 24, 72, 74, 260–261
Hermes, 165, 168, 175
Hesse, Herman, 129
Heyer, Gretchen, 150
Heyong Shen, 31, 185
Hierarchical leadership models, 286–287
Higuchi, Kazuhiko, 61
Hillman, James, 40
Hinduism, 39
Hinton, Ladson, 148, 149
Hogensen, George, 186
Homans, P., 144
Hopper, Earl, 92–93
Horne, Michael, 150
Hoxter, Shirley, 239
Huffington, Clare, 285–286

I

I Ching, 257–258, 260, 265–268
IAAP Russian Revival Project. *See* Russia
Identity. *See also* Cultural identity: of children of transgenerational trauma, 135–136; fear of losing cultural, 23; search for professional, 134–141; shaped by cultural complexes, 118–120
"Illness in the Analyst" (Abramovitch), 57
Immaginando (Chainese and Fontana), 193–194
Imperialism, 291
India, 47–48
Individual. *See also* Individuation: collective harmony and conflicts with, 225–226; collective unconscious and relationship between culture and, 73; cultural complexes of, 75; feeling shame of group, 100; fueled by collective unconscious, 39–40; intertwining political and personal influences in attitudes of, 85; traumas of colonialism to, 39, 47–49
Individuation: of analysts in training from supervisors, 58–59; Chinese philosophy applied to, 262–265; cross-cultural expression of, 10–13;

cultural differences effecting personal, 5–6; deintegration and re-integration, 78; in Eastern and Western cultures, 189–191; establishing boundaries, 46; requiring interaction with other, 40; understanding Jung's views on, 37–39

"Individuation in the Light of Chinese Philosophy" (Beebe), 262

Inferiority, 42–44

Institution as family, 252

International Association for Analytical Psychology (IAAP), 2. *See also* Developing Groups of IAAP; Router Program for IAAP as authority figure, 124; Beebe as liaison to China for, 267; birth of, 25, 67; Russian Revival Project, 162–163, 199, 206–207, 273–291; women's experience training Russian students, 44–45; working in Russia, 79–80

International Conference on Analytical Psychology and Chinese Culture, 262

International Psychoanalytical Association (IPA), 239

Interpreters. *See also* Language: altering nuances in meaning, 165; in conversation with supervisors and supervisees, 163–166; defined, 154; desire for intimacy by, 167–168; emotional involvement of, 166–168; experience of, 165–166, 169–172; giving insights into training, 205; personifying Hermes and Mercury, 163; using dictionaries instead of, 268; working with, 54–55, 201–202, 290

Introversion, 12

J

Jacobi, Jolande, 73–74

James, Thea, 20–21

Japan, 188

Jasiński, Tomasz J., 143–160

Jewish cultural traditions, 229–230

Jung and Eastern Thought (Clarke), 257

Jung, C. G.: collective attitudes of countries, 201; connection with East, 219–220; development of individuation, 11, 12; dialogical writing of, 191; early disciples of, 53–54; Eastern cultural influences on, 21–24; geology of personality, 93; identification with individual and collective, 40; importance of visual imagery to, 186; individuation of plants and humans, 264–265; interest in Chinese culture, 257–258; interest in cultural influences on psyche, 71–72; multicultural insights into work of, 30–32; "Psychological Aspects of the Mother Archetype", 6; questions *I Ching* on how to use it in West, 265; resistance to training institutes, 183; response to formation of IAAP, 67; role of opposites in psychic life, 221; seeing patient as system, 105; silencing of pain by parents, 134; theory of complexes, 74; therapist as participant in individual development, 189; thoughts on collective harmony, 188, 190–191; on understanding meaning, 182; valuing cultural diversity, 4; views on individuation, 37–38; on wholeness, 15

Jung in India (Sengupta), 31

Jung Institute of San Francisco, 260

K

Kachele, Horst, 182, 183
Kalinowska, Malgorzata, 66, 67, 143–160
Kaplinsky, Catherine, 71–89
Karcher, Stephen, 265–266
Kast, Verena, 105, 118–119
Kawai, Hayao, 192
Kawai, Toshio, 189
Kazakhstan, 95–98
Kernberg, Otto, 182–183, 184
Kim, Uichol, 180
Kimbles, Samuel, 24, 74–75, 76
Kirsch, Thomas, 147
Knowledge: embedded vs. explicit, 289; importance of cultural, 26, 28–29; shifting to interactive learning from, 32–33
Kristeva, Julia, 187
Kundalini Yoga, 22

L

LaCapra, Dominick, 184
Lachmann, Frank M., 20
Lakoff, George, 181
Language: common themes in training, 290; communicating with analysands in Taiwan, 230–235; difficulties working with translators, 54–55; Eastern and Western use of, 191; finding sources of misunderstanding in, 222–223; learning with songs, 175–176; as part of supervision process, 80; paternal order dependent on, 186; playing role of symbolic father, 172; role in analysis, 174–175; Router Program training in foreign, 115–116; Russian words for *whisperer*, 202, 205; self-representation through words, 189; sharing silence and speech, 213–215; traveling by learning foreign, 161–162; words that cannot be translated, 61–64
Lao Tzu, 188–189, 258
Leaders: gender roles and women's authority as, 44; hierarchical and distributed roles for, 286; for shuttle training, 285–286
Ledermann, Rushi, 82
Levenson, Edgar, 60
Light: turning around, 263–264
Lithuania, 111–112
Lithuanian Association for Analytical Psychology (LAAP), 113
Liuh, Shiuya Sara, 29, 30–31
Longing to belong, 76–77

M

Mackenzie, 96, 101, 102
Main, Roderick, 265
Malinchismo, 168
Mandala, 15
Mao Tse-tung, 229, 255
Marriage of Heaven and Hell (Blake), 221
Martin, Linda, 20
Martin-Vallas, Francois, 171
Maternal order, 186
Matryoshka, 81–82
Mayer, Elizabeth Lloyd, 186
McGilchrist, Iain, 73
Meaning: interpreter's power to alter, 165; psychoid dimension to, 182
Meier, C. A., 259
Melik-Akhnazarova, Kamala, 161–177
Meme, 73
Memories, Dreams, Reflections (MDR) (Jung), 21, 23
Meregalli, José Francisco, 252
Metaphoric thinking, 181
Mianzi (face), 188–189
Mind: body not letting mind hide

INDEX

feelings, 260–261; children developing own, 237–238, 247; effect of holding children in, 244–245; Jungian theory and Eastern states of, 191–192; not being in anybody's, 247
Minnesota Multiphasic Personality Inventory, 256
Moral complex, 76
Morality and cultural shadow, 75–77
Mother archetypes: cultural influences on, 6–10; Good Great and Terrible, 83; individuation of child and, 11–12
Mother-infant dyads: analysts communications likened to, 230–235; bi-directional influences in, 20; changes in contemporary Russian, 130–131
Mysterium Coniunctionis (Jung), 191

N

Narcissus, 165
Neumann, Erich, 6, 15, 53
Neutrality of interpreters, 166–168
Northoff, Georg, 19

O

Obholzer, Anton, 285
Ogden, Thomas, 211, 289
"Only Breath" (Rumi), 209
Operación Amistad, 238
Organizations, 92–93
Orvydas, Vilius, 117–118
Osterman, Elizabeth, 258–260
Other: analysts seen as, 222; archetypes of universal vs., 221; in Asian culture, 193; boundaries as protection from, 77; confusion and discovery in meeting analyst, 116–118; India as, 48; opening to experience of, 187–188, 209–210; required for individuation and Self, 40–41; revealing shadow in encounters with, 290; stranger as, 124, 222; as virus, 187–188; wariness of, 273–274; working with foreign, 26–27
Outsiders, 291

P

Papadopoulos, Renos, 41
Parents, 134, 136–141
Participation mystique, 23
Paternal order, 186
Patients. *See also* Analysands; Analysts: cultural biases toward countertransference with, 8–9; dynamics of doctor-patient relationships, 94; observing cultural differences of, 5–6; role of Russian analysts with, 45–46, 57–58
Perry, Christopher, 199–208
Philosophy of the Brain (Northoff), 19–20
Pickering, Judith, 135
Pickles, Penny, 275
Pioneers in training, 55–56
Poland, 153–158
Political life, 7–8
Politkovskaya, Anna, 285
Pourtova, Elena, 46, 66, 67
Professional identity in Russia, 134–141
Psychoanalytic training. *See* Training
Psychoid unconscious, 192–193
"Psychological Aspects of the Mother Archetype" (Jung), 6
Psychological Types (Jung), 257
Psychology of Jung, The (Jacobi), 73
Psychology of the Unconscious (Jung), 21–22
Psychotherapy: accommodating cultural differences in, 1–6;

culturally influenced countertransference in, 8

R
Rasche, Jorge, 29
Rasoal, Chato, 28
Ravitz, Liza J., 219–234
Re-integration, 78
Rebeko, Tatiana, 44
Red Book, The (Jung), 22, 30, 221, 262
Redfearn, Joseph, 75
Reeves, Kenneth M., 28
Reich, Kenneth, 166
Relationships: allowing countertransference, 63; Chinese/Taiwanese cultural norms for, 226–227; Confucian principles of behavior in, 190; couple as archetypal pattern, 13; cultural differences in establishing, 45–47; dynamics of doctor-patient, 94; with IAAP officers, 56; individuation requiring, 40–41; between interpreters and supervisors, 168; lines of communication in, 230–231; of Russian practitioners with patients, 45–46, 57–58; universality in human suffering, 228; unlimited by limitations of language, 115–116
Remote training. *See* Shuttle analysis
Reshetnikov, Mikhail, 274
Rezanova, Elena, 167
Rigveda, 21
Ritsema, Rudolf, 265–266
"Rivers of Milk and Honey" (Bortuleva), 66–67
Rosarium philosophorum, 228
Router Program for IAAP: countertransference with routers, 60; expectations of those in, 278; group process in, 98–100, 125; Jung's opinion about, 67–68; origin of term, 55; response of post-Soviet routers to visiting analysts, 59; structure of, 145, 147–148; training in foreign language, 115–116; using cultural sensitivity in, 119–120
Rowland, Susan, 191
Rudakova, Tatiana, 45, 49
Rumi, 209
Running away by street children, 241–244, 249, 250
Russia: adjustment and adaptation in Russian Revival Project, 277–279; cultural complexes observed in, 79–82, 98–100; disorientation of visiting analysts in, 44–47; effectiveness of shuttle analysis, 199–208; hierarchical leadership models in Russian Revival Project, 286–287; history of loss in, 212–213; IAAP women's experience training, 44–45; professional identity in, 134–141; reaction to analysts by analysands, 46, 205–206; relationship of Russian analysts to patients, 45–46, 57–58; Russian Revival Project, 162–163, 199, 206–207, 273–291; significance of Russian nesting doll, 81–82; Soviet psychiatry enforcing political norms, 12; totalitarianism and culture of, 200–201, 284–285; traumas to feminine collective unconscious, 131–133; *whisperers* in, 202, 205; women's gender roles, 128–129
Russian Society for Analytical Psychology (RSAP), 162, 169, 275
Ryan, Joanna, 84–85

S

Samuels, Andrew, 23, 40
Sandplay, 234
Sapir-Whorf hypothesis, 61–62
Scale of Ethnocultural Empathy (SEE), 27–28
Schaverien, Joy, 85
Schellinski, Kristina, 135–136
Schmidt, Martin, 47
Schore, Allan, 292
Schutz, 102
Schwartz-Salant, Nathan, 165
Sebek, Michael, 205–206, 284
2nd European Conference on Analytical Psychology, 29
Secret of the Golden Flower, The, 22, 30, 220, 257, 263, 264
Self: in archetypal patterns, 15; center of consciousness in East, 192; central in analytical psychology, 37; influence of collective harmony on, 188; questions about, 38; requiring interaction with other, 40–41; transcending individual and cultural experiences in, 47–48; turning light of consciousness around in dream state, 264
Self-structures, 149
Sengupta, Sulagna, 31
Sennett, Richard, 289, 291
Separation: at end of Special Time, 251–252; endings and issues of, 65–66; issues in shuttle analysis, 282–283, 290–291; in shuttle analysis, 211–213
Shadow: awareness and integration of, 13–14; boundaries and, 77; defined, 75; linked with identity, 119; morality and cultural, 75–77; revealed in encounters with other, 290; sensitivity in observing cross-culturally, 13–14

Shame, 100
Shearer, Ann, 37–52
Shuttle analysis: authority, teaching styles, and leadership in, 284–288; benefits and limitations of working as outsiders, 291; capacities and limitations of analyst in, 208; common themes found in, 288–290; cultural clashes in, 46, 205–206; developing trust in, 204; differences in boundaries and frames, 279–281; effectiveness of, 199–208, 215–216; encountering hierarchical leadership models in, 286–287; ethical issues in, 206–207; history of, 273–276; Jungian summer school, 283–284; rage of analysands at, 211–213; separation issues in, 282–283, 290–291; transference experienced in, 281–282; transitioning to distributed leadership, 287–288; value of remote training models, 273, 292–293
Sincerely Yours, Shurik (Ulitskaya), 134
Singer, Thomas, 24, 74–75, 76
Skrok, Zdzisław, 143–144
Society of Analytical Psychology, 183, 239
Solzhenitsyn, Alexander, 284
South African cultural complexes, 82–84
Special Time (ST): about, 239–241; examples of, 242–247, 248–251; separation from rounds of, 251–252
"Spirit of Mercurius, The" (Jung), 11
Splendor Solis, 233
Stein, Murray, 1–17, 38, 55
Stevens, Anthony, 76
Stikhi (Bryusov), 127
Stone, Martin, 275

Stranger, 124, 222
Students. *See also* Analysands; Supervisees; Trainees: cultural diversity and supervision of, 8–10; issues of power and authority for, 44–45; pointing out shadow to cross-cultural, 14
Sue, Derald Wing, 20
Superego, 57–58, 76
Superiority, 42–44
Supervisees: acknowledging cultural context of, 61; conversation with supervisors via translators, 163–164; learning about frame for analytic work, 280–281
Supervision: approaches to individuation process in, 11–12; Astor's description of, 144; of children's educators, 240; culturally redefining structure of, 8–9; Jung's role in early, 25; pointing out shadow to cross-cultural students, 14; in Russia, 80–82
Supervisors: Abramovitch's involvement as, 57; acknowledging cultural context of supervisees, 61; avoiding righteousness, 87; conversation with supervisees via translators, 163–164; coping with untranslatable words, 61–64; cultural transferences toward, 59–60; disciple's dilemma for, 57–59; effect of cultural identity of, 100, 101; explaining frame for analytic work, 280–281; feeling culturally disconnected from trainees, 65; identifying Russian cultural complex, 79–82; participating in Special Time for children, 240–241; relationships with interpreters, 168; seeking authority as analyst, 44; understanding cultural coding of archetypal patterns, 61
Survival anxieties of children, 247
Survivor's guilt, 138
Switzerland, 101–107

T

Taiwan: Chinese/Taiwanese cultural relationship norms, 226–227; communicating with analysands in, 230–235; cultural trauma suffered in, 228–229; using sandplay with analysands, 234
Taiwan Institute of Psychotherapy, 185
Tao, 257, 260, 261
Teachers. *See also* Analysts; Supervisors; Trainers: confrontations occurring in Polish training, 151; cultural differences uncovered by, 1–2, 15–16; hierarchical models of Russian Revival Project, 286–287; holding children in mind, 244–245; observing individuation in other cultures, 12–13; pointing out shadow to cross-cultural students, 14; transitioning to distributed leadership, 287–288
Tension in opposites, 221–222
Terrible Mother archetype, 83
"Theory of Liberation" (Freire), 238
Therapy. *See* Analytical psychology
Thoma, Helmut, 182, 183
Totalitarianism, 200–201, 284–285
Trainees: adaptation in Russian Revival Project of, 277–279; confrontations for Polish, 151–152; separation issues for, 282–283, 290–291
Trainers: adaptation in Russian Revival Project of, 277–279; role of, 179–180; working as outsiders, 291

Training. *See also* Router Program for IAAP; Shuttle analysis: Abramovitch's experience in, 54–55; adaptation in Russian Revival Project, 277–279; aims of, 180–182; Astor's description of, 144; authority figures in trainee dreams, 120–123; in China with Beebe via Skype, 267–268; concern about quality and nature of, 288–289; confronting otherness in, 187–188, 209–210; in different cultural contexts, 187–188; disorientation for analysts in, 44–47; effectiveness of shuttle analysis, 199–208, 215–216; emotions of session interpreters in, 166–168; endings in, 65–66; ethical issues in shuttle, 206–207; expectations and tensions in, 58–59; hardest experiences in, 65; initiatory process of, 146, 148–149; Jung's resistance to institutes for, 183; models of, 183, 184, 273, 292–293; personal transformations in Polish, 150–153; prohibitive costs of, 112; restoring femininity to Russian trainees, 133; reviewing remote, 273, 292–293; role of trainer in, 179–180; Russian Revival Project, 162–163, 199, 206–207, 273–291; sharing silence and speech in, 213–215; South African cultural complexes observed in, 82–84; studying Jung in Asia, 185–187; value of shuttle, 288; without institutional containment, 146–147; working with interpreters, 54–55, 201–202

Transference. *See also* Countertransference: between case and group, 103–107; cultural aspects of, 26–27; experienced in shuttle analysis, 281–282; as focus or distraction, 203–204; to pioneers, 56; toward supervisors, 59–60

Translators. *See* Interpreters

Traumas: cultural brutalities to collective, 39, 47–49; cultural complexes developing from, 42; to feminine collective unconscious, 131–133; intergenerational, 134; working transculturally to recover from, 276–277

Tronick, Ed., 20

Trust: developing in shuttle analysis, 204; in interpreter, 202; required in cross-cultural work, 26–27

Tserashchuk, Alena, 44

Tsivinsky, Vladimir, 43, 44–45, 49, 278

Tuckman, Bruce W., 96, 101

Turner, Mark, 181

U

Ulitskaya, Lyudmila, 134

Unconscious: American vs. Chinese Dream in cultural, 223–224; cultural complexes rooted in cultural, 41–42, 72, 74, 192–193; culture and individual in collective, 73; effect of cultural brutalities to collective, 39; effect on group development, 96–98; investigating in partnerships and groups, 93–94; Jung's confrontation with, 23; psychoid, 192–193; traumas to feminine collective, 131–133

United Kingdom, 84–86

Universal archetype, 221

Unus mundus, 191

Upanishads, 21

V

Vanier, Jean, 44
Vassilissa the Beautiful, 129
Vedunya, 128, 133
Visiting analysts. *See* Analysts
Volodina, Elena, 127–133
Von Franz, M.-L., 14

W

Wang Pi, 265, 266
Wang Wei, 263
Western cultures: Eastern cultural complex vs., 219–230; impulse toward individuation in, 11; individuation in Eastern and, 189–191; Jung's view of, 4; questioning dominant assumptions of, 2–4; training models from, 183; Western analyst as other in Eastern cultures, 222
White Flock (Akhmatova), 209
Wholeness archetypal pattern, 15
Wiener, Jan, 29, 44, 45, 46, 273–295
Wilhelm, Richard, 22, 23, 31, 257–258, 260, 265
Wilke, Gerhard, 108
Williams, Gianna, 239, 240, 252
Wittgenstein, Ludwig, 62
Women: authority as leaders, 44; conflict in foregoing collective values in China, 226; effect of trauma to Russian feminine collective unconscious, 131–133; effects of mother's trauma on child's life, 136–141; losing touch with internal woman, 129–131; professional identity of Russian, 127–128; Russian view of sacred femininity of, 128–129
Wright, Susanna, 210

Y

Yama, Megumi, 188
Yu-Wei Wang, 27, 28
Yuan, 189–191

Z

Zinkin, Louis, 40–41
Zoja, Eva Pattis, 31

About the Editors

Catherine Crowther

Catherine Crowther is a Training Analyst for the Society of Analytical Psychology (SAP) in private practice in London. She worked for many years as a psychiatric social worker, family therapist, and psychotherapist in adolescent and adult mental health services in the UK's National Health Service. She was past chair of the SAP's adult analytic Training Committee and is involved in supervision and teaching in the UK and in Eastern Europe. With Jan Wiener, she was joint organiser of the Russian Revival Project, 1996–2010, an IAAP programme for the clinical and academic training of Jungian analysts in Russia. She was also a teacher and supervisor of clinical work in St. Petersburg throughout that time. She has contributed articles and chapters on analytical psychology on subjects such as silence, fairy tales, moments of meeting, cross-cultural exchange, eating disorders, supervision, and training.

Jan Wiener

Jan Wiener is a Training Analyst and Supervisor for the Society of Analytical Psychology in London. She is the former Director of Training. She was Vice-President of the International Association for Analytical Psychol-ogy from 2010–2013 where she was Co-Chair of their Education Committee with particular responsibility for Developing Groups in Analytical Psychology in Eastern Europe. Together with Catherine Crowther, she organised a programme of academic and clinical teaching in Russia and has been involved in teaching and supervising clinical work in St. Petersburg since 1996. She has taught internationally for many years, and is the author of chapters and papers on subjects such as transference, supervision, training, and ethics. She is also author of three books. The most recent, *The Therapeutic Relationship: Transference, Countertransference and the Making of Meaning* was published by Texas A&M University Press in 2009.

For Product Safety Concerns and Information please contact our EU
representative GPSR@taylorandfrancis.com
Taylor & Francis Verlag GmbH, Kaufingerstraße 24, 80331 München, Germany

www.ingramcontent.com/pod-product-compliance
Ingram Content Group UK Ltd.
Pitfield, Milton Keynes, MK11 3LW, UK
UKHW021444080625
459435UK00011B/363